John Horne

A Year in Fiji

Or an Inquiry into the Botanical, Agricultural, and Economical Resources of the

Colony

John Horne

A Year in Fiji
Or an Inquiry into the Botanical, Agricultural, and Economical Resources of the Colony

ISBN/EAN: 9783337063788

Printed in Europe, USA, Canada, Australia, Japan

Cover: Foto ©Andreas Hilbeck / pixelio.de

More available books at **www.hansebooks.com**

OR

AN INQUIRY INTO THE BOTANICAL, AGRICULTURAL, AND ECONOMICAL RESOURCES OF THE COLONY.

BY

JOHN HORNE, F.L.S., &c.,

DIRECTOR OF WOODS AND FORESTS AND BOTANICAL GARDENS, MAURITIUS.

PUBLISHED AT THE REQUEST OF

THE HON. SIR A. H. GORDON, G.C.M.G.,

GOVERNOR OF FIJI, &c.

LONDON:

PRINTED BY GEORGE E. EYRE AND WILLIAM SPOTTISWOODE,
PRINTERS TO THE QUEEN'S MOST EXCELLENT MAJESTY.
FOR HER MAJESTY'S STATIONERY OFFICE.

AND PUBLISHED BY

EDWARD STANFORD, 55, CHARING CROSS, LONDON,

AND

GEORGE ROBERTSON, MELBOURNE AND SYDNEY.

1881.

CHAPTER I.
Description of Tour throughout the different Islands of the Group, &c.—pp. 1-57.

CHAPTER II.
Flora, its peculiarities, &c. Number of Indigenous Species.—pp. 58-73.

CHAPTER III.
Fijian Food Plants. Method of Cultivation. European and Native Vegetables, &c.—pp. 74-92.

CHAPTER IV.
Fruit, kinds of, indigenous to Fiji; and Exotic; might be extensively grown for Exportation.—pp. 93-102.

CHAPTER V.
Native Ornamental Plants. Starch, Spices, Materials for Clothing, Mats Fans, Cordage, &c.—pp. 102-111.

CHAPTER VI.
Indigenous Timber trees. Native Houses, Minor Forest Produce, as Thatch Reeds, &c.—pp. 112-126.

CHAPTER VII.
Conservancy of Forests. Destruction by Fires. Re-Foresting and Preservation of Water Supply.—pp. 127-137.

CHAPTER VIII.
Botanic Garden. Industrial School. Museum. Meteorology.—pp. 138-148.

CHAPTER IX.
Number of Islands in the Group. Area of Land. Reefs. Rivers. Navigation. Watersheds. Mountains. Lakes. Taviuni. Rabi. Loma Loma. Koro. Ovalau. Levuka.—pp. 149-162.

CHAPTER X.
Hot Springs. Rocks. Minerals. Soil, &c.—pp. 163-170.

CHAPTER XI.
Agricultural Products. Copra. Coir Fibre. Cotton. Sugar. Tobacco. Cacoa or Chocolate. Tea. Vanilla. Coffee. Cinchona Bark. Rice, &c.—pp. 171-184.

Q 2019.   Wt. P 456.

CHAPTER XII.

Labourers. Markets for Produce. Land Titles. Kinds of Native Produce, &c.—pp. 185-191.

CHAPTER XIII.

Stock. Pigs. Fowls. Fauna of Fiji.—pp. 192-194.

APPENDIX I.

Report on the Caoutchouc-yielding Plants of Fiji.—pp. 195-202.

APPENDIX II.

Report on Sandalwood Preservation, and extension by Cultivation, &c.—pp. 203-212.

APPENDIX III.

Propositions regarding a Forest Ordinance for Fiji.—pp. 213-235.

APPENDIX IV.

Letter to the Hon. the Colonial Secretary, Fiji, and Rules suggested to regulate the Felling of Timber Trees in Forest Reserves. —pp. 236-246.

APPENDIX V.

Meteorological statements.—pp. 247-255.

APPENDIX VI.

List of Flowering Plants and Ferns Indigenous to Fiji.—pp. 256-286.

NOTE.

In pronouncing and spelling Fijian words the following rules are observed, viz. :—

B, b, pronounced *Mb*, as Bau, *Mbau*; Bua, *Mbua*.

C, c, pronounced *th*, as Cibicibi, *Thimbithimbi*; ca (bad) *Tha*.

D, d, pronounced *Nd*, as Druadrua, *Ndruandrua*; Kadavu *Kandavu*.

G, g, *ng* in sing or rang as Turaga, *Turanga*; vuga, vunga.

Q, q, *ng-ga* or *nka*; as Waga (canoe) *Wangga* or *Wanka*.

PREFACE.

In 1877, I accepted an invitation from Sir Arthur Gordon to visit the Islands of Fiji, of which he was then Governor; and availing myself of the privilege of leave of absence granted to Civil Servants, I left, *en route* for England, viâ New South Wales, Fiji, and San Francisco, after a resident service in Mauritius of 16 years.

Sugar being the staple product of that Island, the introduction of new varieties of sugar-cane and their culture had received my special attention for a number of years; and, on leaving, the Chamber of Agriculture, with the sanction of Government, commissioned me to select and forward whatever new and suitable specimens of canes I could find in the different islands lying on my route.

I spent a year in Fiji, and my notes of travel during that period are given in the following pages.

JOHN HORNE,
Director of the Botanic Gardens, and of
Woods and Forests, Mauritius.

November 4th, 1880.

CHAPTER I.

DESCRIPTION OF TOUR THROUGHOUT THE DIFFERENT ISLANDS IN THE GROUP, &c.

My route through the Fiji islands is traced in red on the annexed map, and I now propose giving a description of it and of the country through which I passed.

Before starting, His Excellency the Governor of Fiji, the Honourable Sir A. H. Gordon, C.B., G.C.M.G., furnished me with a circular letter of introduction written in Fijian to all the chiefs, informing them who I was, the object and nature of my visit, and requesting them to give me all the assistance in their power in providing me with carriers, guides, &c. This they most cheerfully and willingly attended to, and I found no difficulty in moving about. In each village some one, generally the schoolmaster, " teacher " or native clergyman, was found who could read and explain the letter to the people, who were at all times attentive listeners. This was a surprise to me; but I found in the course of my journeys, that many of the grown-up people had been taught by the Wesleyan missionaries; that a church and a school,— the latter of which in small villages was a place of worship as well on Sundays as on other special occasions,—were established in every native town— (Koro). These were well attended, as most of the rising generation could read, write, and cipher to some extent. In most native houses worship was conducted night and morning, showing that the missionaries had nobly done the work of Him who sent them. The teachers were supported by the voluntary contributions of the people. I had with me an intelligent half-caste boy (Rae Spowart), about 15 years old, to act as interpreter. I took with me a mosquito net, rug, mats for sleeping upon, sugar, tea, coffee, biscuits and bundles of paper for drying plants, all enveloped in waterproof coverings; and, thus equipped, I traversed the country with my native guides and carriers.

The first journey was to Lavoni, in the centre of the island of Ovalau, over rugged and steep paths. On the way through the forest a good collection of plants was made. The soil was seen to be rich and fertile, and the rocks volcanic breccia or agglomerate. The Lavoni valley,—in which I spent a week, traversing it in all directions, making fine collections of its flora—seen from an eminence is not the least beautiful of the many pretty places in Viti. Its extent is about 7 square miles. It is well watered, and the soil is rich and suitable for sugar cane and coffee cultivation. My route was down the valley to Bureta by a native track, which took me repeatedly across a considerable stream, and sometimes compelled me to wade in water up to my knees. The soil, consisting of alluvial deposits, was rich and fertile, and densely covered with the wild sugar cane (vico), reeds, &c. At Bureta I went to the native church and heard a sermon from a Fijian. He preached extempore from a few notes, was earnest if not eloquent, and seemed neither to want ideas nor words in which to express them. On Sundays such preachers may be heard in every village from one end of Fiji to the other. At

Bureta there is a considerable tract of fine flat land well adapted for sugar cane cultivation. From Bureta I crossed to Moturiki, a small island with an area of 8 square miles of fine land, which is well suited for the growth of cocoa-nuts. From Moturiki I visited Viti Levu and entered the Rewa river by its eastern mouth, and went up to na Koro Vatu, about 65 or 70 miles from the sea. I was everywhere delightfully surprised with the fertility of the land, the size of the river, the fine scenery, and the luxuriance of the cultivated sugar canes, which, to a small but increasing extent, are grown by a few settlers on the alluvial flats bordering the river. All the kinds of sugar canes, their healthiness, their capability of resisting drought, and of "ratooning" after cutting, as well as the quantity of sugar and weight of cane per acre which each kind yielded, were subjects of earnest inquiry; as on the eve of my departure from Mauritius for the South Seas, I was instructed by the Government of that colony, at the request of the planters, to collect and despatch thither all the different varieties I could obtain, together with useful information respecting the peculiarities of each kind. In Fiji alone I obtained over twenty different kinds, samples of which were sent to Mauritius in 16 wardian cases. On the banks of the Rewa and its affluent streams, including the deltas formed by its several mouths, there are about 400 square miles of land unrivalled in quality, and especially well adapted for growing cane. Most of this land has deep water frontage. A large proportion of it belongs to settlers, some of whom cultivate sugar canes, and others do not because they cannot get their canes crushed, the sugar mills being either insufficient or at too great a distance. I also saw on the Rewa two thriving cotton plantations, each about 200 acres in extent, belonging to the natives. They were in good condition, well laid out, cleanly kept, and the plants had a healthy appearance. The centre road through one of them was bordered with banana trees for shade; and these trees were also grown to mark off the plantation into squares or plots, each of which belonged to a different township. At the junction of the Wai ni Mala and the Wai ni Buka, about 60 miles from the sea, I saw a plantation of tapioca or cassava (*jatropha manihot*) and arrowroot, and a few acres of a young flourishing coffee plantation. I noticed during my journey that all the rocks on the banks of the river were of the sedimentary formation.

On my way back to Levuka I visited the Wesleyan Training School at Navaloa, which is attended by advanced scholars from all part of Fiji, either for the purpose of being educated as teachers or for the native ministry.* It is also attended by the sons of the chiefs for the purpose of obtaining a higher education than the village schools can supply. The establishment is under the superintendence of a master who is a European.

On returning to Levuka I went by sea to Tai Levu, a large district on the east coast of Viti Levu. We were becalmed for some time at the "back (N.W. side) of Ovalau," and, on landing, slept at the house of a settler who is rearing a fine herd of cattle there. The quality of his land was good and capable of growing sugar canes; but here again the want of a mill and the want of means to purchase one were hinderances in the way of cultivation. These wants indeed are common

---

* *Vide* page 140.

to all parts of the group, even in places where extensive areas of cane land exist. We left this settlement in the morning, and in the afternoon reached our destination, proceeding up a deep but narrow creek shaded by *tiri* (mangrove) trees, whose overhanging branches had been cut to allow the tall mast of the boat to pass. At this part, and on the banks of the Wai Delici (pron. Wai Ndelithi) the land is amazingly fertile, and formerly produced good crops of cotton. Near the river there are extensive alluvial flats, but the general character of this part of the country is steeply sloping and gently undulating low hills and valleys. This land had been cultivated by the Fijians, but now, being either not required by them to grow " food" upon, or more likely, owing to their method of cultivation, it has become densely covered with wild sugar canes, reeds, grass, and such hardy kinds of trees and shrubs as can best resist the fires which periodically burn up the dry herbage. An opinion prevails among the settlers, that sugar canes will not grow on the low hills so common in this and other parts of Fiji. What I saw here, on the Rewa and at many other places in the group shows that opinion to be an erroneous one. Here the canes were growing and in good healthy condition on the sides and tops of the low hills where cotton would not thrive. However, in these places their growth is less rapid and the canes a little smaller than on the fat alluvium on the banks of the rivers. Still they do grow well and prove remunerative. Of course such land will not be much in demand for cane cultivation until all the rich bottom lands are fully occupied.

In this locality I visited another cotton plantation owned by natives. Like those already mentioned it was in good order. When on the Rewa my attention was called to the unhealthy condition of some cocoa-nut trees, and here I noticed the injury was caused by an insect eating the under side of the leaves, which then become covered with brown spots and wither and decay (if the ravage have been extensive) frequently before being fully expanded. From the leaves being thus injured or destroyed, the trees are unable to bear fruit, and sometimes are entirely killed. In addition to about 30 square miles of good cane land, there is also in this locality a large area well adapted for the growth of the cacoa (chocolate) and Liberian coffee trees, both of which delight in a moist warm climate and thrive at a low elevation above the sea.

I next went to Rabi (Rambi), an island which lies off the north-east point of Vanua Levu and between it and the northern end of Taviuna. It contains an area of about 28 square miles. The soil is rich and fertile. The island is well wooded and well watered, and the prevailing rock is agglomerate; but both aqueous and basaltic rocks were frequently seen. It is also mountainous, and some of the valleys are of great picturesque beauty. Cocoa-nuts are extensively grown for copra, which is here dried by artificial means, viz., by heated air. Coir fibre is extracted from the husk of the cocoa-nuts by steam machinery. The cocoa-nut plantations have been carried up to a considerable elevation on the hill sides, on which they are healthy and thriving. An idea is entertained by many of the settlers that the cocoa-nut tree will not thrive on the sides and tops of the hills near the sea and within the full influence of its breezes. This notion is wrong, and disproved by what any visitor cannot help seeing,—thousands of cocoa-nut trees growing not only on the sides of the steep hills, but even on the tops of the ridges. In these places the growth of the tree is slower and the produce a little less than on the rich low-lying land on the shore. There are about 6 square miles of good cane land on Rabi, and a large portion of the island is suitable for both Liberian and common coffee, leaving a wide margin round its shores for cocoa-nuts, and a large space on the tops of the mountains for timber to attract moisture. From the position of this island the majority of the plants growing upon it are also common to the adjacent portions of the two large islands near it. As every facility for collecting and drying plants was put at my disposal by the kind proprietor, I thoroughly explored the island, a large portion of which is still covered with primeval forest. Here I made collections of over 300 species of flowering plants and ferns. These I left at Rabi during a six weeks' tour in the northern parts of Vanua Levu. This enabled me to go quickly over a portion of that large island, as I had gathered on Rabi a great number of the plants which grow on the parts of Vanua Levu lying nearest to it. I also thus saved the transport on men's shoulders of bulky packages for upwards of 200 miles, as well as avoided delays in collecting, *en route*, the daily shifting of numerous specimens and the drying of the paper from which these specimens had been removed. Besides, by acting as above, I did not bring from a distance specimens of plants which I could obtain at what, for a time, was my head quarters.

From Rabi I crossed to Koro-i-vono, a native town on the eastern side of Vanua Levu. At this place there is a considerable area of good cane land, and plenty of space on the beach for extending cocoa-nut plantations. Here a deep stream enters the sea, up which boats drawing 6 feet of water may proceed at high tide. From this I decided to travel to Savu-savu by Natawa bay. The path, a well-kept one for a short distance, led up the coast for about 4 miles through groves of cocoa-nut, bread fruit, and other trees; it then, as a mere track, broke off to the left through the forests, ascended the mountains along the beds of streams, over rocks, down the sides of precipices to the district of Togaloa, on the south-east shore of Natawa bay. In passing through the forests which densely cover the mountains lying between Koro-i-vono and Natawa bay, many fine specimens of the *dakua*, the Fijian *kauri* pine, *dakua—salu—salu*, and other kinds of good timber trees were seen for the first time, and several kinds of ferns were added to my collection. The *dakua* formerly abounded at Koro-i-vono,—Seemann mentioning that some large ones existed there in 1860-61;—but European sawyers have played sad havoc among them, and large trees were only found in the more inaccessible parts of the mountains.

At Togaloa I learnt that all the Bulis or district chiefs in the province of Cakaudrove had gone to Somo-somo, in the island of Taviuni, to attend the provincial council, which the Roko Tui, or supreme chief of the province holds twice in the year. The Fijians have (and before the islands were ceded to Great Britain had) an elaborate system of polity by which all tribal and provincial matters are discussed and regulated. As I could not get carriers from one district to another without coming, in some degree, in contact with this system, an outline of it may be interesting.

The starting point in this system is the village or "Koro." Over every village there is a local chief called the Turaga ni Koro. Over one or many villages, or perhaps over a district, there is a Buli; and over these is the chief of the province, or Roko Tui. The Governor, as Her Majesty's representative, is the supreme chief of all. These native functionaries, although the office which they hold is hereditary, *i.e.*, belongs to certain families, are elected to office by the district council, of which more anon. The village chief is assisted by a council of elders, which meets once a week, and executive officers, magistrate (Turaga ni lawa), policemen, town-crier, inspectors of gardens, &c., are appointed to carry out its decisions. These act on behalf of the community as guardians of the peace, see that villages are kept clean, that fences are not broken down, that animals are not destroying the gardens, and as messengers, guides, &c. The affairs of the district are regulated by the district council (Bose ni Tikini), which consists of the Buli and all the village chiefs of the district. This council, which meets once a month, nominates all the village chiefs--whom it may suspend if not dismiss from office; discusses and regulates all local matters, such as the opening of new roads and the repairing of those already made, making and repairing bridges,—allotting at the same time the portion of the work which each village has to do; keeping bathing places in decent order, cleaning villages and superintending the payments of the village officers out of the local rates. The Bulis of a province must meet twice a year in council (Bose vaka Yasana), and discuss the affairs of the province with the Roko Tui. These (there are 12 of them) with two Bulis chosen from each province, and the native stipendiary magistrates meet once a year in the great council (Bose vaka Turaga), and discuss the " native affairs of the nation." At this meeting each Roko gives a detailed report of the province of which he has charge. These reports are severely criticised by the other chiefs, and suggestions are offered as to such executive and legislative measures as the assembly would like to see adopted by the Government. The resolutions of the great council are mere recommendations which the Government of the colony is free to accept or reject.

On arriving at a village I was conducted to the Bure ni Sa, or strangers' house, one of which is to be found in every village, or to the house of the chief. The Governor's letter was read, and the Turaga ni Koro in council appointed the guide and the men who were to be the carriers to the next town or district, and without such not a man would lift a package. The state of the native roads, the order and cleanliness of a town were soon seen to be sure indications as to the character of a Turaga ni Koro. Whatever might be the decisions of the village council, it was his particular duty to have them carried out.

Koro ni Saca was the next halting place, where I stayed two days, making excursions into the forests, collecting plants and examining the country. At this town a large stream enters the sea, and boats drawing 8 feet of water can go up for 3 or 4 miles into the interior. In the locality there are about 6 square miles of land suitable for growing sugar cane, lying on the banks of the river and on the sides of low hills. The mountains are well wooded with valuable kinds of timber trees. The soil is suitable for growing coffee. The road from Togaloa, a well-made and well-kept path, runs along the beach, shaded by fine *ivi*, *dilo*, and *vulu* trees, through several villages which nestle among groves of cocoa-nut and bread-fruit trees. Numerous clear streams are crossed, and native plantations of cotton, yams, and *masi*, or native bark-cloth trees, are seen. Some of these plantations are in good order, but the condition of others leaves much to be desired. On the beach and on the low hills cocoa-nut trees could be largely planted. At present, in these places there is little but useless scrub, tall grass, reeds, &c. In the vicinity of Natawa village, there are several settlers, and about 3 square miles of what appeared to be good cane land. Between Koro ni Saca and Viene the hills end abruptly on the shore, forming in many places cliffs of agglomerate and basaltic rocks, which extend quite into the sea. The path between the towns was stony and hilly, but the land was fit for growing cocoa-nuts, occasional groves of which in hollows were passed. I intended to spend the night at Viene, but on arriving I found the people were about to send a canoe to Vuni Sawani, a village at the head of Natawa bay, and about 7 miles further on the journey to Savu-savu. On learning that the path leading to that town was a bad one, I gladly accepted the offer of a passage in the canoe.

Viene is situated on the side of a point of land which projects into the bay. A small but high lying island lies at a short distance from the mainland, from which it is separated by a channel that is sometimes dry at low water. The land in the vicinity of the village is well adapted for growing sugar canes, of which I saw a native plantation of about 10 acres to supply thatch,

and canes for eating. The Fijians are very fond of
the sugar cane, and never seem tired of eating it—
sucking its sweet juice and throwing the pith away.
When going from place to place, they frequently
carry large quantities of it to use on the way, in
order to beguile the tedium of the journey, and to
give as a present to friends and acquaintances. For
such purposes it is largely used on festive occasions,
and the quantity of cane which a Fijian will consume in an idle hour, or while listening to a story,
astonished me.

The canoe was soon got under weigh; the large triangular shaped sail was hoisted, and the breeze being
favourable and not much sea, the 7 miles were run
over in about 40 minutes. On the way I noticed
several settlers' houses, native villages, small groves of
cocoa-nut trees, and land cleared for young plantations
at various places along the shore. At this part of
the island the land is narrow. It is in reality an
isthmus, only about 3 miles broad from Natawa
bay to the main ocean on the south-western side of
Vanua Levu. The character of the country is hilly.
The low hills are densely covered with long grass and
occasional patches of forest, but the soil is good, and
cocoa-nuts could be more extensively planted there
than they are. Some of the sheltered valleys are
also suitable for coffee, the Liberian species would
most likely answer best, and in a few places good
sugar cane could be grown.

Vuni Sawani is a low lying place not more than
2 feet above high water. The village is built upon
land, which had recently been a *tiri*, or mangrove
swamp; a place where one might fairly expect to find
malarial fever prevalent, did such exist at all in Fiji.

Numerous villages on the coasts in all parts of Fiji are built in similar situations. It is not too cleanly kept, and the pigs roam at will through it,—the inhabitants putting up low fences in front of their doors to keep these animals out of their dwelling-houses. With the exception of the village constable and the garden inspector, or guardian, all the men belonging to the village were at Wai Levu, helping the Buli to make *taro* patches, and were not to return for a few days. However, the women, to show that their town should not be wanting in the accustomed hospitality to strangers, volunteered, in the absence of the men, to carry my packages to the next town, a distance of 2 or 3 miles towards the interior, where I could get men to carry them on to Savu-savu.

At Vuni Sawani there are about 7 square miles of superior cane land consisting of alluvial flats and low hills, and sufficient space for enlarging the cocoa-nut plantations, which in this locality are not so extensive as they might be. The inhabitants of this village have extensive *taro* patches, which they cultivate with great skill and care. The soil of the locality,—a brown heavy loam, appears to favour the growth of that plant, and the beds, " patches," of it which I saw were in fine condition, equal if not superior to any that I saw in any other islands of the Pacific ocean. A canal averaging $2\frac{1}{2}$ feet in breadth, and $3\frac{1}{2}$ feet in depth, and more than 2 miles long, conveys the water to these patches from a neighbouring stream. I started with a guide early in the morning, and after walking leisurely for about 3 hours, was overtaken by the women and boys with the packages just as I was about to enter the next town. I paid them and gave each a present of tobacco (for all the Fijian women smoke), for which they were thankful.

It being mid-day when I arrived at the village, and finding all the men were at work making a new road, I left my packages to be sent after me the next morning, and then, with a guide, pushed on to Savu-savu, where I arrived at 4 p.m. The distance from Savu-savu bay to Vuni Sawani is said to be only 7 miles; but by the zigzag way I was taken by the guide from village to village, it was not less than 16 miles. This is *vaka viti, i.e.*, Fiji custom or fashion. Some fine views of the country were obtained, and several fertile valleys were passed where coffee, cacoa, &c., could be grown. About 4 square miles of fine cane land, consisting of flats and low hills, were observed,—notably in the vicinity of Savu-savu bay, though separated from it by a ridge of low hills. The general character of the country is hilly, steep, and undulating; but the quality of the soil is fair and good, and water abounds in all the valleys. Most of the land had been cultivated, and afterwards abandoned, in accordance with the custom of the people, and allowed to become overgrown with rank grass and scrub. The latter appeared to be prevented from growing and spreading by the fires which annually burn the grass. The rocks were sedimentary, agglomerate, and some columnar basalt was noticed. From experience now gained in travelling in Fiji, in order to avoid delay and annoyance, I henceforth endeavoured to engage the carriers to the village which was to be the end of my journey for the day, as on arriving at a village, situated about half way, the men were absent; engaged in their plantations or elsewhere, which naturally enough they did not like to leave, and the men who came with me just as reasonably objected to carry further than they had agreed to, even when the offer of double pay was made to them.

They would, however, have cheerfully gone beyond the half-way village had they before starting understood they were required to do so.

At Savu-savu there are extensive boiling springs, which phenomenon, at another place, will be more particularly alluded to. The scenery at this place is beautiful and grand,—lofty hills, cloud-capped mountains densely covered with trees, deep dark looking valleys, contracting almost into gorges,—low grass-covered hills which in some places advance and present precipitous terminations to the sea whilst in others they retreat from it, leaving flat beaches between their base and the shore,—groves of cocoa nut, bread fruit, *ivi, dilo, vesi,* and other trees, all surround a bay which is 12 miles broad at the bottom, 5 at the head, and from 6 to 7 miles deep. In the bay there is a small island 150 feet in height, on the top of which stands a house. This bay is one of the finest and largest natural harbours in Fiji; it has deep water close to the shore, is easy of access, whilst the reefs prevent the swell of the ocean from entering it. Several large streams run into Savu-savu bay, which are navigable for large boats of light draught for some distance.

It is a favourable place for a township; and, although it has been decided to build the capital of Fiji at Suva in Viti Levu, a large town, from a variety of circumstances, will in all probability be built upon the shores of Savu-savu bay. The two highest parts of the small island in the bay are united at the base by a mangrove swamp, in the centre of which there is a deep pool, probably the crater of an extinct volcano, into which there is a channel for boats. The pool, I have been told on good authority, would make an excellent dry dock, as it only requires the mud to be dredged out to enable it to receive large vessels. Several of the settlers in this vicinity have herds of fine cattle. Cocoa-nut plantations could be largely extended on the low hills that in some places border the bay.

A kind friend sent me across the bay in his boat, which landed me at Wai-wai, near Wai Levu, on its northern shore. Here an enterprising young settler has bought land, on which he has built a small sugar mill, and is growing cane and coffee, the former on the rich alluvium at the base of the hills, and the latter in the sheltered and fertile valleys. Cocoa-nut trees abound on the shore, but there is room for more extensive plantations of this valuable tree. The low lying undulating country extending from this place to the Bua mountains, a distance of 50 miles, is well watered, and suited both by climate and soil for growing sugar cane, cacoa, Liberian coffee, &c., with a broad fringe of cocoa-nut trees along the coast. Tracts of this fine land belong to settlers, many of whom are too poor to do anything with it, even if they were willing and had the necessary knowledge and skill. The greater part of it, however, belongs to natives, who annually clear portions of it on which to grow their food plants, allowing the portions already used to relapse into jungle.

I stayed a week in this locality, arranging the specimens collected on the way from Rabi, and making excursions into the mountain forests. I may remark that the farther I proceeded from that island the more numerous did plants, which I had not observed upon it, become. I also found that the specimens already collected were drying badly,—becoming mouldy and the leaves dropping off on account of the heat and moisture, notwithstanding the changing of the drying paper twice daily. Specimens will not dry properly when made into bundles and enveloped in waterproof coverings; yet without these coverings the paper carried exposed through this damp country would become a mass of pulp in a few hours. The carriers were careless; indeed no ordinary care could prevent the packages from coming constantly in contact with the tall grasses and branches of shrubs dripping with wet, that border the narrow tracks leading from one village to another. Three changes of drying paper were in use, but from my constant movements it could not be thoroughly dried, although well aired, spread out and hung up in the interior of native houses every night. Henceforth, with the exception of specimens of tender ferns, all other specimens were well withered in the sun and air before they were put into drying paper. To do this, they were carried in open baskets by day and hung near the fires and round the sides of the *bure ni sa* at night. From the specimens being scattered at night and exposed to the wind by day I lost about 150 numbers (representing as many species of which I had memoranda) out of a total of 1,150 kinds gathered.

My next journey was to Naduri, in the province of Macuata, on the north-west coast of Vanua Levu. Walking slowly, collecting all the way and stopping a night at Loma loma, a village in the centre of the island, Naduri was reached on the evening of the second day. On the chart the direct distance from Wai-wai to Naduri is 18 or 20 miles, but, by the indirect route of the native paths, it is fully 35 miles, On the first day my course lay through the mountains. The path, rough and apparently not much used, ran along streams, up steep ascents and down awkward descents, over slippery boulders and fallen trees, up the sides and along the crests of mountains. These mountains are densely wooded, whilst the soil in the valleys is fertile and well adapted for growing coffee. The head waters of the Drekiti river—here a large, swift, running stream—were crossed on the first day, and I gathered some interesting and new species of flowering plants and ferns in the mountain forests, which are here well stocked with fine timber trees. Among these were noticed the *dakua, dakua-salu-salu, kausia, kau-tabu, lewininini, damanu,* &c. Along the paths I remarked that the Fijians have at intervals recognised halting places, from which all the scrub has been cleared, and where they light fires, cook their food, rest, and smoke. These spots are always on the tops of ridges, commanding good views of the country. The air on the top of the mountains felt fresh and cool; but the climate is exceedingly wet, and a fair day quite the exception. In such a climate ferns, &c., abound, and the branches of the trees and the petioles of the leaves are covered with mosses. On the second day, the route was through a country of an almost opposite character. The climate here was moderately dry. The hills, consisting of calcareous and other sedimentary rocks, were low and undulating, and covered with grass, patches of forest, rough ferns, and the turmeric and *yabia* (Fijian arrowroot) plants. The character of the plants changed, there being but few of those I had seen in the wet districts. Fires which break out annually among the grass do a great amount of injury to this province. In the bottoms and on the sides of some of the valleys there is a large amount of good fertile land capable of growing sugar cane, and in other places large areas of pasture land.

Cocoa-nut trees, loaded with fruit, are common, but not numerous in the vicinity of the village. The inhabitants of this part seem to be poor; their villages are by no means clean, and, altogether, compare unfavourably with those residing on the coast. On the road to Naduri four large streams were crossed, all flowing westward to the Drekiti river. There are several settlements on this river where, I believe, there are nearly 30 square miles of good cane land, most of which contists of alluvium and low hills with water frontage on a navigable stream.

The village of Naduri is built on a low lying, flat piece of rich land about 3 square miles in extent, which is capable of yielding any kind of crop. The climate being dry during the cool season, cultivation is aided by irrigation. The water is supplied by one or more streams that have been turned into channels dug for the purpose. Cocoa-nut trees abound in the village and in the vicinity, but there is available space for further planting. The village is well built, cleanly kept, and the ground is carpeted with short grass like a lawn,—the latter feature being characteristic of the tidiest and best kept native villages. The European stipendiary magistrate for Vanua Levu resides and has a court house in Naduri. All the magistrates in Fiji, Levuka excepted, go from one district to another administering justice, holding their courts, for want of proper buildings, in the village schools and occasionally under a tree. But this inconvenience is in the way of being remedied. In the villages where courts are usually held the people are erecting buildings for the accommodation of the court, and also for the magistrate where it is necessary for him to remain a few days. Bèche de mer is found on the reefs off this coast, and the natives are largely employed in getting and curing it for the market. Sandalwood used to grow in the forests of the province, but owing to reckless felling, it has almost disappeared from them.

The interpreter being unable to walk far on account of a sore foot, the Roko of Macuata sent us up the coast in a *takia*, or dug out canoe, to Labasine, opposite Mali island. Thickets of mangroves fringed the shore and were occasionally backed by small grooves of cocoanuts, for extending which there seemed to be ample space. The hills descend in places to the water's edge, and here and there form precipitous cliffs of agglomerate. The surface is covered with rank grass and dotted with the *balawa* (Pandanus, or screw pine) trees and spots of forest. Near Mali three or four navigable streams enter the sea, and the mangrove swamps are extensive. There are a few settlers near these streams, on the banks of which I should think a large area of cane land existed.

A change of men at Labasine to manage the canoe was not one for the better. The canoe had not been long out when the men, from the noise of the surf breaking on the barrier reefs and other appearances, predicted a storm and advised landing at the village of Wavu wavu. A severe storm of lightning, thunder, wind and rain came on as predicted and lasted the whole night.

The next morning I travelled to the village of Vuni vutu, from whence the Turaga ni Koro took me to Tutu in his large sea canoe. Tutu is nearly opposite the island of Draudrau. The coast from Labasine is a series of bold, projecting bluffs of agglomerate, interspersed with seams of coralline sandstone, enclosing a succession of bays varying in depth from 2 to 3 miles; and in breadth from 3 to 4. At their heads these bays terminate in mangrove swamps, behind which are small groves of cocoa-nuts. One or more streams enter the sea at the head of each bay, and they are navigable for small vessels for several miles inland. These streams having neither bridge nor ferry, travelling along this part of the coast can only be done by good swimmers,—even they would have to encounter the danger arising from the numerous sharks which infest all these waters. The ridges that extend seaward are thickly covered with *noko noko* (casuarina) and *balawa* trees, brushwood and grass. Low rolling hills lie beyond the flats, and like those just referred to are covered with grass and brushwood. In making their roads the natives have, in several places, dug deep into these hills, and have exposed the sub-soil, which is found at times to consist of white and red clay. The estimated area of good cane land on the Macuata coast, extending from Macuata island to Udu Point, is about 40 square miles. Most of it lies in small areas, of 100 acres and upwards, each. In working such land small and comparatively inexpensive "plant," capable of making 2 or 3 tons of sugar per week, will be particularly useful in developing this and similar localities in Fiji. On the other hand, many districts will give employment (during the proper seasons of the year for cutting and crushing the sugar cane) to numerous mills, each able to make 20 tons of sugar per day. At Tutu a few strange plants were found among the volcanic and sedimentary rocks which surround that place.

I left Tutu late in the afternoon in a large canoe and went up the Kuru-kuru river for about 8 miles; then landed and walked 7 miles to the village of Kali kosa. As it was dark when we landed the natives made torches of dry bamboos, which burnt well, to guide us down the steep banks and over the fords of two large streams. On the banks of the Kuru-kuru river there is a large area of good cane land. In a small lake near Kali-kosa there is a floating island, in connexion with which the Fijians have some long mythological stories to tell.

Next day the route was in a south-westerly direction to Nadoga, which was reached about 4 p.m. It was a wet walk all along the banks of the Wai ni Koro, a deep and rapid stream, which was crossed and re-crossed 15 times. The natives are extremely kind, and at the fords insist upon carrying the *papalagai*, *i.e.*, the foreigner, across. One soon gets tired of this, besides there is the risk of the bearer slipping, and bringing both down to be carried away by the current. One gets equally tired of changing clothes when fords are frequent, and the result is you plunge into the water with your ordinary costume upon you, even though you may find it up to your shoulders. A considerable extent of fine flat land was noticed on the banks of the stream, and in general the country consisted of low-lying hills covered with grass and brushwood and pasture land. The rocks noticed were principally sedimentary and volcanic, and some had the appearance of decomposed granite. The sanitary condition of Nadoga leaves much to be desired. The mosquitoes were more numerous and troublesome than in any other part of the group. I left on the following morning, when the path led through the mountains, up the bed of one stream and down the bed of another to Malaka, or Malata, a village on the north-west shore of Natawa bay. During this day's journey the land was found to be good throughout, and in the mountains the greater part of it was well wooded and suitable for coffee growing. On approaching Natawa bay the path led through some fine cane land about 2 square miles in extent. In the mountains which extend from one extremity to the other of Vanua Levu, a distance of 120 miles, there is an area of about 700 square miles of prime land well adapted by climate and soil for growing the coffee tree.

The following day my journey lay for 15 miles along the shore of Natawa bay to Vatu Kura. It passed through numerous cleanly-kept villages, which were well shaded with groves of cocoa-nut, bread-fruit and other trees. The path—a well-made one—lay along the beach, occasionally on flat ground, but not unfrequently cut into the sides of the hills, where it was supported on the outside by rough walls of stone. Many streams—flowing from the mountains—were crossed, some of them by bridges of single, slippery, crooked, logs, trunks of trees worn smooth, and others of more architectural pretensions, all more or less dangerous to walk upon. Near Vatu Kura we were ferried over a large stream, one at a time, in all that remained of a canoe that had been recently broken. It was simply the half of a canoe, the water being prevented from entering at the broken end by a wall of clay. The character of the country on this side of the bay is similar to that on the south-east side already described. Many young plantations of cocoa-nut trees were noticed, and there is ample vacant space for more. Most of the low hills are covered with grass and "bush," and the soil is very fertile. Basaltic, agglomerate, limestone, and other rocks of aqueous character are very conspicuous.

At Vatu Kura the signal agreed upon,—smoke if in the daytime and fire at night,—was made to my friends at Rabi, who at once sent a boat for me, and I arrived at that island after a somewhat stormy passage of eight hours. In a few days, the steamer, which trades in the group and carries the mails between the several islands, called at Rabi, and I returned to Levuka, via Laucala, Loma loma, Mango, Kanacea, Vuna Point in Taviuni, and Koro.

The first place the steamer stopped at was Wai ni Buli, Tasman's straits, on the northern part of Taviuni, where we remained all night. As steaming among the islands at night is dangerous, and at places even by day without a pilot at the mast-head to point out the sunken rocks, the vessel has to anchor in a safe harbour, or lie to at a fair distance from the shore, every night. We steamed across the straits at daylight, and soon anchored at Luacala (pr. Lauthala). This part of Fiji is exquisitely beautiful. The water is deep, smooth, and blue; and sunken rocks and detached pieces of coral are readily distinguished by their colour. Numerous islands rise suddenly from the sea, and huge masses of rock are seen towering up to a height of 800 feet, clothed with dense forest, and surmounted by the tall trunks and waving leaves of the cocoa-nut palm and tree fern. Quame (pr. Kaima), is a small, well timbered, high lying, picturesque island, on which many cocoa-nut trees may be seen nestling at the head of little coves, and growing far up on the rocky heights wherever the soil is rich enough for them to grow. Laucala is a fine, fertile island, nearly the whole of which is capable of being made into a grove of cocoa-nut trees. That tree is well represented here; indeed, those that I saw were among the healthiest in

Fiji, but they were growing too closely together. The steamer went close to Nataimba, a good view of which we had in passing. It contains about 9 square miles of good land, and nearly the whole of it is planted with cocoa-nut trees. After entering inside the reef at the northern end of Loma Loma or Vanua Balawu (Exploring islands), the steamer passes within 20 yards of the shore, along which the scenery is most charming. The island is well wooded at this part, and rises abruptly from the sea to an elevation of several hundred feet. The shore is well protected from the force of the waves by the barrier reef. But the action of the water upon the coral limestone rock has worn it away, so that at one uniform level all along the shore the upper part projects at a little above high-water mark. There are many islets protecting half hidden coves or small bays, in some of which there is deep water. The steamer remained nearly 24 hours at the village of Loma Loma. There are numerous settlers at this place, and it is a port for both inward and outward bound foreign vessels. Some healthy young mahogany and toon "cedar" trees, and good crotons from the New Hebrides and New Ireland were seen in the gardens.

Maafu, the Roko of the Lau province, has granted the use of a few acres of land, which the inhabitants of the village have laid out as a public botanic garden. Of course, I paid a visit to it, and was much pleased with what I saw. It is highly creditable to the settlers and natives, and says much for the energy and zeal of both, for I believe the latter take a lively interest in " their garden."

There are many Tonga men in this province, and its Roko is the nephew of King George of Tonga. The

Tonguese are a fairer race of men than the Fijians, and both men and women are of larger build. They readily combine to carry out any enterprise. This was shown in the wars between the two races. Individually they are acknowledged to be less industrious and not such good handicraftsmen as the Fijians; they are also more haughty in their bearing towards strangers. The houses here have circular ends, a feature I did not observe in any other part of the group.

Copra is the great article of produce of this province, and there is vacant space for the cocoa-nut tree being extended a hundred fold. The island is long and narrow, thinly wooded near the village, and its rocks are agglomerate, basaltic, and aqueous. The village is built in a grove of cocoa-nut and bread-fruit trees. It is cleanly kept, well laid out, and divided by broad paths, which serve for streets, and which are covered with short lawn-like grass.

The next port of call was the island of Mango, rendered famous from the cotton grown there having gained the highest awards at the International Exhibitions held at Philadelphia and Paris. It is a fertile well-cultivated island, about 9 square miles in extent. The attention of its owners has been attracted to raising coffee instead of cotton, which latter does not now pay so well as formerly.

From Mango the steamer went to Kanacea, where it shipped a quantity of cotton. This had to be conveyed nearly 2 miles in small boats through a narrow passage in the reef, which is here 2 miles broad and the same distance from the shore. The island belongs to settlers, and contains extensive groves of both old and young cocoa-nut trees. The other places touched at were Vuna Point and Koro.

My next trip was to Suva, in the island of Viti Levu. This is the site of the future capital of Fiji. In many respects a better could not be chosen. The scenery of the locality is among the finest in Fiji. The harbour is large and commodious, easily entered, and well protected by the reefs, whilst the water is moderately deep. From this harbour, and inside the reefs, there is water communication on one side to the mouth of the Rewa river, and by that stream into the interior of Viti Levu ; and on the other side for 30 miles along the coast and up the Navua river. Three streams enter the harbour on its northern side, navigable for small boats for several miles. A road, a portion of which is already blocked out, is to be constructed to the upper Rewa and the Wai Manu districts, where some of the most fertile land in Fiji is to be found. The site is on a projecting point, which on the land side rises to an elevation of 300 or 400 feet. On the other three sides it is surrounded by the sea and open to the breezes,—the S.E. trades, which keep the locality fresh and cool. The elevated land, as well as a large space on the northern shore of the harbour, which can be reached by small boats in any weather, is suitable for the erection of villas. As indicated on the plan of the town, I believe the streets are intended to be half a chain (33 feet) wide. This for a town in the tropics, where a good supply of cool pure air is absolutely essential, seems to me to be too little, at least for the main streets. These should not be less than 50 feet wide. They may even be wider to allow for side walks, each from 10 to 12 feet broad, and for the planting of a row of evergreen trees on each side for shade. The ground is naturally well drained. The underlying rock, which is but a short distance below the surface, resembles marl.

From Suva I made several excursions into the surrounding country. Most of the land in the vicinity belongs to settlers. In some parts cotton had been grown, in others sugar cane, but both these have been abandoned for grazing herds of cattle. At Suva I discovered the insect, a small moth, which in its caterpillar stage of development eats the leaves of the cocoa-nut trees. One of my journeys was to the upper Rewa, *viâ* the native village of Kaluba, returning by the sources of the Tamavua river. Several others were made, both by land and sea to na Vesi saw-mills, where I obtained much information respecting Fijian timber from the intelligent proprietor. On the way to Rewa the country is densely covered with reeds and scrub, which are periodically devastated by fires. The land appeared to have been cultivated by the natives. In many places it is very fertile. A young but promising coffee plantation was shown me, in which the plants looked healthy and thriving. On the banks of the Wai Manu, an affluent of the Rewa, there is an extensive area of excellent cane land. The greater part of it belongs to settlers, some of whom have fields of fine healthy looking sugar cane, and others have herds of well-fed cattle. The land which lies between the two rivers is also very good. This had been previously cultivated by the Fijians; but it is now covered with a thick young forest. Forests of this kind, which spring up where land has been once cultivated, are numerous in Fiji. These, together with the many abandoned *dalo* or *taro* patches, lead to the belief that the country was once much more populous than it is now. But the clearing and cultivation of land, followed by abandonment, may be rather due to the habits of the people than suggestive of a diminished population, though it is true that epidemics of measles have carried off a very considerable number. On returning to Suva the Wai Manu was crossed 4 miles higher up, and the route followed was that which had been surveyed for the road above mentioned. We passed the sources of the Tamavua river, and the village of Colo ni Suva. Near this village can be obtained a fine view of the lower reaches of the Reiwa, extending to Ovalau, and also some fine mountain and river scenery. The soil along this route is fertile, and in many places suitable for growing sugar canes, whilst in others coffee would be found to be more suitable. The rocks are principally of a calcareous nature.

When leaving Suva I was joined by Mr. Langton, who travelled with me through the interior of Viti Levu examining sites for the natives for their coffee plantations. We proceeded first in a boat for about 30 miles along the coast, and then went up the Navua river. At the mouth of this river, where we remained a few days examining the country and visiting some of the settlers, there are about 60 square miles of prime cane land, consisting principally of low hills and alluvial flats. Cotton was once grown here, but since that ceased to pay, the settlers have turned their attention to raising stock. As in other places the want of sugar mills prevents the cultivation of the sugar cane from being entered upon.

At the village of na Quave we met the Roko of the province (Roko Tui Namosi), who had assembled a number of natives,—men, women, and boys, to clean a cotton plantation belonging to the tribe. The plants looked healthy, but thinning had been neglected when they were young. The yield of cotton would also have been greater if the plants had been cut down after bearing the first crop. We left na Quave for Nukusari, a village about half-way to Namosi. For 14 miles the road led through magnificent forests, in which there is a large amount of excellent timber, such as *dakua, dakua-salu-salu, kau-tabua, damanu*, &c., I made considerable additions of rare species to my collection of plants.

Although our course now lay for some distance along an elevated ridge, it was so shaded by trees that, with the exception of a magnificent view of Koro Loa, a good view of the country could not be obtained. Koro Loa is a lofty, conical shaped, rugged mountain, which rises to an elevation of about 2,000 feet above the sea. The land surrounding it appeared to be low lying hills and valleys, but so densely covered with forest that the surface could nowhere be distinctly seen. The soil was everywhere rich and fertile, and the rocks seen were principally basalt and agglomerate, and occasionally some of a sedimentary character. Before arriving at the bed of the Navua river, the path suddenly descended the bed of a stream with narrow, lofty, perpendicular banks so overgrown with bamboos and wild canes that no view in front could be obtained. When these were pushed aside, we turned sharply round a rock, and to our surprise found we were standing on the edge of a river (a branch of the Navua), 10 yards broad and $2\frac{1}{2}$ feet deep. We forded this stream three times in the short distance of 500 yards. Then climbing its right bank where it joins the Navua river, we crossed a projecting point of land about 6 yards broad, and descending into the Navua, which we forded on a spit of sand and gravel a yard in breadth at the top, and with $3\frac{1}{2}$ feet of water upon it. This spit is thrown across the Navua by the current of water which its tributary brings down and which at this point enters the main stream at right angles. It was remarked at this place, that the channels of both rivers had been worn down through a coarse grained sedimentary rock which rose in steep crags to a height of 300 feet above the water. Here the Navua is from 25 to 30 yards in width, and at the head of the rapids, where we forded it, the water was sometimes 3 feet deep and very clear. The route now followed the bed of the river, crossing it three times within the next three miles. This was necessary in order to avoid climbing steep banks at places where the river by taking a sudden bend strikes into their sides, leaving no space for a path between them and the water.

Nukusari is a small village on the left bank of the river. By the people of the Namosi tribe it is used as a resting place between the coast and the interior. We left it next morning, and fording the Navua and a tributary stream five times, we arrived at Koro Wai-wai about 4 p.m. At Nakusari I had a slight attack of Mauritius fever during the night, and next day in fording the river for the first time, I felt quite giddy when in the middle of the stream, but calling the guide to take my hand I arrived safely on the other side. The water was nearly 3 feet deep at this place and running rapidly, and the bed of the river was covered with loose stones which frequently rolled from under the feet. The depth of the water and the strength of the current rendered it dangerous to lift the feet; but with the aid of a stick and pushing one's foot forward a few inches at a time, the river was safely forded.

Many of the natives were employed taking yams to the coast on rafts made of bamboos. It was an interesting sight to see, from a height, long trains of these rafts floating down the river one after another. A man on each raft guides it with a long pole, avoiding shoals and rocks, keeping it in deep water and preventing it from running against the banks. The land between Nukusari and Koro Wai-wai is of excellent quality and well suited for growing coffee. Much of it had been cultivated, but had been allowed to relapse into forest.

From Koro Wai-wai excursions were made up a branch of the Navua river, to the villages of Lasalasi and Biba. The land seen was fertile and well adapted for growing coffee. As seen from the cliffs on the banks of the river, the rocks were agglomerate and sedimentary. We left Koro Wai-wai and after a short journey of about 5 miles arrived at Namosi. This village lies in the bottom of a beautiful valley surrounded by lofty mountains, some of which (Voma) rise to an elevation of nearly 3,000 feet above the sea. The entrance to the valley and the exit from it are on the course of the Wai-dina. At the village this river is about 6 yards wide, and the surrounding scenery is indescribably fine. For many reasons the site of Namosi is well adapted for an inland town. The climate is delightfully cool at night and not too hot during the day. There are no mosquitoes, at least in the cool season, and then the temperature at night falls to 55° and even 50° F., whilst in the day it rises to 70° and 75°. We stayed a week at Namosi, and from it took long rambles into the mountains, which are well wooded with many fine specimens of the best kinds of timber trees peculiar to Fiji. A young, petty chief acted as our guide, who, on being asked what countryman he was, drew himself up and, with a great assumption of dignity, replied, "Kai Biretania Saka," a British subject, Sir. We ascended Mount Voma, from the top of which an extensive view of the country, indeed of the greater part of Fiji, was obtained. It extended to Ovalau on the east and to Kadavu in the south; on the west and north, a dense forest and a series of hills rising above and beyond each other until they seemed to meet in the distant horizon. The view was an extensive one, but many of the objects were too distant to stand out in anything like distinctness. The prettiest objects were the valley and the windings of the Wai-dina, which latter, like a silver thread, meandered through all the varied shades of colour which a fertile valley, partially wooded, partially cultivated, and well populated, ever presents to the eye. Namosi used to be a clean and well-kept village, but lately it has been much neglected. In times gone by some of its people and their chiefs were notorious cannibals. The stones, which are said to be a record of the number of bodies consumed by the chiefs, may be seen standing near the site of the old heathen temple, or "Devil's Bure." The rocks in the vicinity are basalt, agglomerate, and aqueous, but principally the two former. The land is very fertile; in the bottom of the valley and on the sides of the low hills surrounding it, sugar cane might be grown, and coffee in more elevated parts.

From Namosi we travelled to Vienunga (as pronounced), a village about 20 miles to the west. When about 4 miles on our journey, we were shown the place where a shaddock tree once grew, of which it has been said, that the fruit which fell on one side was carried to the sea by the Wai-dina and the

Rewa, and that which fell on the other side by the Navua river. We passed Koro Wai-wai on the way, and although the path was on the top of an elevated ridge the view to be obtained was an extremely limited one, on account of the rain and fog. The path afterwards became very bad, precipitous, muddy, and slippery. Several streams were passed, the path taking us occasionally along their rocky beds. After crossing the Wai ni Awa, at a cleanly kept village of the same name, the path ran along the crest of a ridge whose rocky sides were almost perpendicular, and large rocks seemed to block the way. We scrambled over these huge rocks by the aid of rough steps cut in the slippery stone and upright logs with notches cut into them for the feet. The top of the rock was a plateau on which several circular terraces, one above another, in the form of fortifications, had been made by the natives. The place has evidently been one of the old "fighting towns," or forts, into which the people retired in troublous times, when not only tribes had their feuds, but neighbouring villages even seemed always to have a standing quarrel to be fought out as opportunity offered. These old fortifications are numerous in Fiji. Every mountain top is such a one, and the narrow mountain ridges leading to it are cut across by deep ditches to act as lines of defence against invaders, and to render the "fighting town" safer to dwell in. Hence, in a great measure, the reason is to be found why the Fijians directed their paths along the beds of streams and the tops of ridges, from the latter of which good views of the country could be obtained. These made progress difficult, and, for the same reason, the immediate approaches to villages were often of the most intricate kind. The position chosen for a fort was naturally a strong one, and, where art was called in for its aid, it was easily made impregnable to the modes of warfare practised by the old Fijians. They were generally well supplied with provisions, and also with water from frequent showers. Within a few of them there were springs of water and even ground for growing bananas,—plantations of which still remain. For the last 7 miles to Vienunga, where we arrived about 5 p.m., the path was a well made one and led through virgin forests. The country passed through from Koro Wai-wai was very hilly. Rocks of a sedimentary character were most numerous, but volcanic breccia and basalt were also occasionally seen. Up to our arrival in the vicinity of Vienunga most of the land had been cultivated at one time, and then, in accordance with the custom of the country, allowed to rest and to become overgrown with scrub and reeds. The soil was seen to be of good quality and suitable for growing coffee and tea.

Vienunga is a large and clean village on the banks of a small river which falls into the Navua. As a rule the villages in Fiji are built on the banks of rivers and large streams; always near water. This village is situated in the centre of the forest which extends for miles round it in all directions. A portion of this forest is felled annually, and the land turned into plantations for growing *yams, dalo, bananas,* &c. We stayed here several days and had long rambles in the forests, in which several new plants were found. These forests are well stocked with timber trees, especially the *damanu,* which is one of the best and most useful woods to be found in Fiji. It is a pity to see so much timber wasted here; fine trees are felled and then burnt off the land, while large imports of timber for building are constantly arriving from New Zealand and Oregon, and yet the *damanu* is superior both to the *kauri* of the one country and to the *pine* of the other. The timber cut down for clearing might, in many instances, be floated down the rivers to the coast, where it could be utilised. This should be done after the timber is seasoned in the log, as some of the kinds, when green, are too heavy for floating, or, their buoyancy might be aided by the use of bamboo rafts.

A large number of the natives in Namosi, and some other villages which I visited, are covered with what is locally known as the Tokalu, or Solomon Island ringworm. This is, I believe, a fungus, which lives and spreads in the skin until it extends all over the body. The skin becomes rough, and changes from its natural dark colour to grey, and the body of the infected person at a short distance appears as if covered with scales, like a fish. The disease, if it may be so called, is infectious but not painful, and it does not seem to impair the strength or energy of those affected by it. The Chief Medical officer of the colony has prescribed an effectual cure for it, but it is difficult, if not impossible, to get the natives to use it long enough to make a cure. At Vienunga the natives have a cure of their own, the sap of a tree which they call *kau karo*. The tree abounds in the forests in different parts of Fiji, but I was not fortunate enough to get either the flower or the fruit of it. It seemed to be a species of Oncocarpus. One or more species of this genus are indigenous to the islands, and the juice is acrid and burning. The natives use the juice by allowing a few drops to fall on the parts affected, which it inflames and burns out. The Fijians are careful not to allow any of it to touch their fingers, but I tried some on the back of my hand without noticing any injury done.

We visited the nurseries of young coffee plants which the Government are getting up to aid the natives in establishing coffee plantations. At Namosi and other villages which we passed, the seedlings looked well and thriving, but nowhere had they been so well attended to as at Vienunga, and at no place were they to be found in better health. Vienunga was the first village where we found the language to differ from that spoken on the coast. Very few of the common people understood the interpreter. As we were to visit many other places in the interior, where the coast language is only spoken by a few, I engaged a young man who was familiar with both, and by his help and our own interpreter, we were soon able to carry on all necessary communication.

The journey from Vienunga to na Moali was, for the fourth part of the distance, through the forest. The path was very rough; fallen trees blocked the way, and at almost every few yards a stream had to be crossed; then it conducted us along the bed of the Wai Moali, a small river which flows into the Navua; indeed, all the streams crossed by us since we left Namosi are tributaries of that river. The land which we passed through was fertile, and both it and the climate were suitable for coffee, tea, or cinchona cultivation. The rocks seen were principally of aqueous origin, but those of an igneous nature were abundantly represented. The village of na Maoli lies in the region between the wet and dry parts of Viti Levu. Open grass-covered hills, which are interspersed park-like with clumps of forest, are characteristic of this locality. The soil is fertile, and the surface is not too steep for growing the sugar cane. Tea and coffee would also succeed here. Agglomerate rocks abound, but some of aqueous nature were also noticed.

After passing the village of Vosi Dam the path to na Tua-tua-coko, or Fort Carnarvon, was a very rough one indeed. For about 9 miles it went through innumerable beds of *dalo*, along irrigating water-courses, and then down the rocky bed of a mountain torrent to Wai Basaga, on the left bank of the Siga Toka river. The forests which lie between Vosi Dam and na Moali are full of the best and largest timber trees that I have seen in Fiji, and a very large proportion of them have reached a mature stage of growth. At Vosi Dam an extensive view of the province of Navosa was obtained. This is the dry part of Fiji, and the chief features in the scenery are low hills covered with grass and small tracts of forest in some of the valleys and on a few of the mountains. We forded the Siga Toka at the village of Wai Basaga, and after crossing some low hills covered with grass and recrossing the river, we arrived at Fort Carnarvon, so called after the late Secretary of State for the Colonies. The country passed through consisted of hills with very steep sides, which had been much disfigured by land-slips. Many *dalo* plantations were seen in terraces on the hill sides. The annual fires cause great damage to this part of the country. The greater part of the land is of fair quality, but it suffers periodically from drought. Sedimentary and basaltic rocks were the most common kinds observed. The houses in this part are built in a different manner from those on the coast. Their length is about equal to their breadth, and they have high pitched roofs. The thatch is thickest at the eaves, where it is bent into a horizontal position and then cut short. Fort Carnarvon was built by the Government after the late rebellion was quelled. Its site is on the left bank of the Siga Toga river, about 250 yards from the stream. It is built on the top of a low hill surrounded by plains, and is occupied by from 60 to 80 men of the armed constabulary, all of whom are natives drawn from the various provinces of the group. They are under the command of a native lieutenant, who is the son of a renowned Fiji chief,— Ritova of Macuta. The Governor's Commissioner, who also acts as magistrate of the province, has his head-quarters at the fort. Of all occupations the natives like soldiering the best, and when on duty they have a dignified and martial appearance. Up to a certain point they learn their drill very rapidly, and go through the movements with great precision. A native chaplain is stationed at the fort, who also acts as teacher to the men.

Besides attending to their military duties, the men are employed in planting and raising food for their own consumption. Their food a short time ago was all purchased from the people of the vicinity; now the greater part is grown by them on the flat land in the neighbourhood. Built in the centre of the district in which the late rebellion broke out, the fort was intended to be of service in any similar outbreak,—an event not likely again to occur, as the people all seem cheerfully to have accepted Government control, and are quite ready to submit their differences to its decision. This part of the country is occupied by numerous petty tribes, who were almost constantly at war with each other, except when they patched up their differences to settle any difficulty they might have with their neighbours on the coast.

From the fort I went to Pickering's Peak (Koroba) to see the sandalwood which was said to grow on its sides, returning by way of Wai-wai, and the bridle path between Nadi and Fort Carnarvon. The country here was very hilly and the land covered with long grass and clumps of the *vunga* tree, metrosideros polymorpha, &c. On the tops of the mountains and in many of the valleys there are patches of forest. In these a large amount of excellent timber and other useful trees, the *lauci*, aleurites triloba, are found, a portion of which is annually cleared by the Fijians to make room for their food plantations, or destroyed by raging fires. The land was from fair to good in quality. In many places it is suitable for coffee, and in some others, as at na Sau-coka, for sugar cane. In the vicinity of Fort Carnarvon I visited some large caves among the calcareous rocks there, which were once a stronghold of the cannibals. The entrance to the caves was defended by walls and a fence of lemon trees, on which I found some human bones which the inhabitants of the caves had exposed to view to show all who passed what might be the result of a quarrel with them. This custom was not uncommon among the Fijians when they were cannibals, the shaddock being the favourite tree on which to expose the bones of their victims. These bones were generally put in the forks of the branches, or where the branches united with the trunk. If the trees were young and thriving, the bones in a few years became embedded in the wood. I could not learn why the shaddock, lemon, and lime trees were selected for this purpose, and presume it was simply because these kinds abounded in the vicinity of villages to a much greater extent than others.

In the dry season the Siga Toka river, at Fort Carnarvon, is about 30 yards wide, and from $1\frac{1}{2}$ feet to 2 feet deep. As far into the mountains as na Tua-tua-coka there is a large area of excellent flat land which is suitable for growing sugar cane. Near the coast also the area of such land is said to be extensive, and I believe that, with the low hills and the alluvial flats, there are from 20 to 25 square miles of good cane land on the borders of this river. The locality is dry, but land which has such a quantity of fresh water running through it to the sea, should not be much affected by drought. Coral limestone and a rough calcareous looking sandstone abound in this neighbourhood, but volcanic breccia and basaltic rocks are not uncommon.

We left Fort Carnarvon for Nadrau, stopping the first day on account of a violent thunderstorm at a village called Mata-wala, the next at another called Bilo, and arriving at Nadrau on the third day. The path was in some places along the beds of streams, then it ran along the tops of the hills, occasionally crossing inverted V shaped ridges at right angles; over rocks and down precipices at angles varying from 80° to 90.° It was annoying, after a long climb up one side of a steep hill, to see the path winding among the long grass, up and over a similar hill just before us, and also to know that we had to go down an almost perpendicular descent of nearly a 1,000 feet, to leap from one slippery stone to another along the rocky bed of a stream in the bottom of the

valley, and again to ascend to an equal if not greater
elevation than the one we were standing upon. We
passed through numerous small tracts of forest, all
of which seemed to have sprung up after land had
been cultivated. These were interspersed with open
pieces of land thickly covered with reeds.. Although
we traversed this district in the dry season,—and the
locality is one of the driest in Fiji,—yet, judging from
the vegetation, the rainfall at some period of the year
must be considerable, probably amounting to from
80 to 90 inches per annum. I noticed on this journey,
and also in the vicinity of Pickering's Peak, that in
passing over the wooded mountain tops, showers and
fogs constantly occurred, and the vegetation was
dripping with moisture, while on the open grass land
at some distance from the woods, and at about the
same elevation, the dust was blowing on the paths, and
and the grass was perfectly dry. The streams
which were passed during this journey were affluents
of the Ba river, the sources of which we had traversed
until we arrived at the top of the ravine or gorge at
the bottom of which the village of Nadrau is situated.
Rocks of sedimentary origin were the most abundant
kinds; but volcanic breccia in the beds of a few streams
and also at Bilo was well represented. Basalt was
not uncommon, notably at a village about midway
between Nadrau and Bilo. The strata of the sedimentary rocks on the tops of the ridges were generally
lying in a horizontal position, and those on the sides
were more or less inclined. Several varieties of sedimentary rocks abound in the province of Navosa, and
marble is said to have been discovered there. Near
the village of Wai-wai these rocks form cliffs about

100 feet in height. Some of them are coraline, and at the caves alluded to I saw pieces of coral embedded in them. Most of the land passed through was fertile, and in many parts suited for coffee growing. Good pasture abounds everywhere, and this part of the country is well adapted for rearing cattle, horses, and sheep.

An old heathen temple exists at Nadrau, the only one I saw in Fiji, but it is far gone to ruin. We stayed at this place a day or two and then went on to na Babuca, on the Wai ni Loa (or Black river), a branch of the Wai ni Mala. After crossing the Siga Toka we ascended the cliff on its east side by a narrow path. We subsequently proceeded up a narrow valley, the path running along the bed and by the side of a canal, made with some ingenuity and much labour by the natives on the steep, rocky side of the valley to irrigate *dalo* beds. About a mile or so in a straight line from Nadrau we came upon some marshes among which the Wai ni Mala has its course. Then on we passed through some flat park-like country, and entered the primeval forest near the village of Wai Dra-dra. At this village the inhabitants of Nadrau grow their provisions, as *yams, bananas, land dalo,* &c. As at Vienunga, a portion of the forest is annually felled for the purpose of making new plantations. The journey hence to Babuca was the most pleasant that I made while in Fiji. The path was through the virgin forests, well shaded from the sun, either level or gradually descending. Here I made a good collection of rare ferns and other plants. The timber trees usually seen in such localities abounded. On the way we passed some agglomerate and basaltic rocks, but the sedimentary kinds were most common. The soil was of excellent quality, and suited for growing either coffee, cinchona, or tea, the two latter especially in the neighbourhood of Wai Dra-dra.

Babuca is the name of a small district as well as of a village. We stayed in it several days, and had long rambles in the forests. This part of the country is exceedingly mountainous, and is well wooded. Portions of the woods are annually cleared as elsewhere; but on account of the superabundance of rain, nature soon restores the balance by the rapid growth of other portions; and, also, on account of the moisture, fires are neither so extensive nor destructive as in the province of Navosa. Agglomerate is the kind of rock that most abounds, but those of aqueous origin were also met with. The land is fertile, and coffee would thrive well upon it.

On leaving Babuca the path to na Koro Vatu led down the bed of the Wai ni Loa to its junction with the Wai ni Mala, and afterwards along the bed of that river, nearly as far as the village of Ruku-ruku. We found it a rough one, and, following the course of the river, we had to climb over boulders, and to cross the river 15 times. The depth of water at the fords below the confluence of the two rivers, varied from 3 to 5 feet. The scenery of the river was extremely pretty. The country was mountainous and well wooded. At several places the Wai ni Loa has cut through inverted V or triangular shaped ridges, exposing whole series of strata. On the journey to na Koro Vatu we stopped a night at Koro Suli, a large village on the left bank of the river. A short ramble into the forest was made at this place, when a new species of gardenia and several species of ferns were collected. On resuming our journey the country began to open out, to be less densely wooded, and less mountainous. The course of the river is crooked. The large flats on its banks are eminently suited for the growth of the sugar cane. In the vicinity of Ruku-ruku and of na Koro Vatu there is an area of about 50 square miles of good cane land, consisting of these alluvial flats and low hills. Near these two villages, and also near Kami Rusai, beautiful and extensive views of different parts of the country were obtained—mountains, valleys, wood, water, picturesque looking villages, and cultivated land. Na Koro Vatu is a large well-kept village, situated, as its name implies, on a rock, which overhangs the Wai ni Mala. In this neighbourhood large nurseries of coffee plants are being formed under the watchful eye of the intelligent district commissioner. With the young plants it is intended to establish plantations to be cultivated by the natives who live in the mountain districts near the sources of the Wai ni Buka and Wai ni Mala. In the deep pools of the last named river sharks abound, and a short time before my visit a child, who had accidentally fallen into one of these pools at na Koro Vatu, was devoured by one of these creatures within sight of its mother, while she was occupied in washing her cooking utensils on the edge of the water. I also heard the case of a man, who about the same time went alone to bathe in one of these deep pools, and was severely bitten by one of these voracious animals—the flesh being torn from one of his thighs. We forded the Wai ni Mala, about a mile below na Koro Vatu, and travelled across the country to na Bukè Lukè, on the Wai Dina. At that place the Wai ni Mala is about 40 yards broad, 3 feet deep, and the current runs from 2 to 3 miles an hour. It is subject to sudden floods, and sometimes rises 17 feet in a single night. The country traversed in this part was covered with low hills, on which grass and trees were growing, the land having been previously cultivated. The soil was fertile, and adapted for growing Liberian coffee, cacoa, and in some places sugar cane. The latter could be grown more particularly towards the course of the Wai-dina, on the banks of which there is a large area of flats and other lands on which it could be grown to perfection. I stayed a few days at na Bukè Lukè and ascended Bukè Levu, a mountain about 1,800 feet above the sea level, from which fine and extensive views of the country could be obtained. Na Bukè Lukè is about 20 miles east of Namosi by the course of the river. The rocks seen on the journey were mostly sedimentary; on the sides of Bukè Levu agglomerate, basaltic, and aqueous rocks were common.

I left na Bukè Lukè in the morning, and arrived at Veseri village about five in the afternoon. The path led through the forest, in which I gathered several new species of ferns, and made additions to my collection of flowering plants. The forest was well stocked with good timber, such as *damanu*, &c. The path was steep and wet, especially the latter part of it, which took us down the bed of the Veseri river, after having crossed the Wai Manu and several tributary streams. The land which we passed over was good and favourable for growing cacoa, and both the common and the Liberian species of coffee. The rocks were mostly aqueous, but both basalt and agglomerate were noticed at different places, particularly in the bed of the Veseri river.

Veseri is a small village situated on the river of the same name, about 3 miles towards the interior from the head of Suva bay. The river is navigable to the village for boats drawing a few feet of water. Near the village there are nearly 3 square miles of flat land and low hills on which sugar cane could be profitably grown. My friends at Suva on hearing that I had arrived at Veseri, came in a boat for me at midnight and took me to their home, where I once more experienced their hospitality and kindness. Two days after my return to Suva the steamer formerly alluded to called, on her monthly trip, and I returned to Levuka *viâ* the western and northern parts of Viti Levu.

On the way a short stay was made off the mouth of the Navua, where a mail was landed, and then proceeding past Serua, we anchored for the night in Nadroga harbour. The next place of call was off the Tavu river, up which, with a small party, I went 8 miles in a rowing boat. Here we saw a fine plantation of tobacco belonging to an enterprising settler, in whose garden we saw some rare exotic trees which, with much care and at great expense, had been introduced from Australia, and they seemed to be doing well. On the river there are about 6 square miles of fine cane land. The country in the vicinity is hilly, covered with long grass and dotted with screw pine trees. The soil is not bad, but it has been much injured by the fires which periodically burn up the grass. We next stopped at Nadi, where there are numerous settlers, who, instead of growing cotton as formerly, now give their attention to the cultivation of maize. Some of these settlers are raising stock, and getting up herds of Angora goats, for which the place seems well suited. The sugar cane would flourish well here, but there are no mills to crush it or to make sugar. Off the mouth of the Ba river the steamer stopped for about an hour, then proceeded to Nananu Islands and

Viti Levu bay, anchored for the night, and arrived at Levuka the next morning.

The northern parts of Viti Levu are very mountainous. These mountains are covered with grass, and here and there with screw pine trees, standing singly or in small clumps. Patches of forest were seen on the tops of a few hills, and in the valleys between them. Fringes of mangroves were noticed on the shore, and the large mud flats at the mouth of the Ba river were covered with them. Occasional groves of cocoa-nut trees were seen growing behind the mangroves, and settlers' houses and native villages were frequently noticed. At Nananu an energetic and enterprising gentleman is breeding Angora goats, rearing silkworms, and has a promising plantation of young cocoa-nut trees. At Viti Levu bay there is a colony of settlers who are mostly growing maize. The steamer leaving at daylight, I had no opportunity of landing at this place, but I believe there is a large area of flat land both here and on the Ba river, as well as in other parts of the northern coast of Viti Levu, extending from Nadi to Tova peak. From reliable information on this subject, I estimate the area at about 60 square miles. It is good cane land, and I have no doubt that coffee would grow in many parts in the mountains at a short distance from the coast. Good pasturage abounds all along this coast, on which sheep, horses, and cattle could be reared in numbers more than sufficient to meet the requirements of the group.

From Levuka I went in the steamer on its next trip to Bua. This is a large province or district in the south-west of Vanua Levu. The sandalwood, for which this district was once famous, has almost disappeared, only a few trees remaining. Bua is said to be one of the driest parts in Fiji, but according to the observations made by Mr. Holmes at Delanasau in this district, the rainfall varies from 80 to 159 inches in the year, the average for the last six years being 118 inches.

I remained here for a week in the hospitable home of one of the settlers, during which time I made several journeys to what still remained of the forests, collecting about a hundred different species of flowering plants and ferns which I had not met with in other parts of the group. At Kadi there is a thriving coffee plantation which is being gradually extended. There are about 15 square miles of alluvial land—river flats, and low hills, which would grow sugar cane well, and still leave space to increase the cocoa-nut trees nearly a thousand-fold. The country is covered with long grass and reeds and pieces of forest, especially in the ravines. Fires in this, as in so many other parts of Fiji, do great injury to vegetation and to the soil. One of the settlers is getting up a herd of fine cattle, which will help to keep down the rank grass, and so tend to prevent fires from spreading. The river here is navigable for several miles inland, and the alluvial flats bordering it are very extensive. I left Bua and journeyed to Wai-Nunu, across the southern point of the island. On the way I stopped for a night at the village of Warei. The path was in some places rough and wet, and in the latter part of its course it went along the banks and bed of the Wai Levu to Warei. In every district of Fiji there is a Wai Levu, *i.e.*, a large river or stream. In most parts of the country passed through, the soil was good, and suitable for growing the common coffee and that of Liberia. The mountain, valley, and water scenery at Warei was very fine indeed. The inhabitants of this village were the poorest I had seen in the group, and their houses were in a very dilapidated condition. The burying ground, as at many other places in Fiji, was close to the village. In many instances the Fijians attend most carefully to the graves of their departed friends. These resting places are commonly parterres planted with choice and gay flowering plants, ornamental trees and shrubs. Not unfrequently the graves are covered with mosaic-like patterns made of variously coloured pieces of coral and pebbles, gathered one by one at immense labour, and during, perhaps, many years. From Warei I went to Koro Levu, a village on the south-east coast of the island. The route for some distance was through the mountains and then along the coast, the path being well made, but in many places steep. The soil in the mountain district was of good quality and capable of growing coffee. Near the coast it was rich, and sugar cane, cacoa, and Liberian coffee would thrive well upon it. There is ample space on the coast for an indefinite extension of the cocoa-nut tree.

At Koro Levu the Commissioner of Lands was holding his court,—conducting a patient and painstaking inquiry into the claims of settlers to lands in the district purchased from the natives previous to the ceding of the islands to Great Britain,—"Crown grants" or title deeds being given in all cases where satisfactory proof could be shown.

The rocks noticed were principally aqueous and agglomerate, but some of a basaltic character were also seen. There now being several large estuaries to cross between Koro Levu and Wai-Nunu, I was advised to take a canoe, and on arriving at the latter place I found a cutter, which a much respected gentleman had sent from Savu-savu bay to take me to Taviuni. At Wai-Nunu the land is fertile, suitable near the sea for cocoa-nuts (for which there is space for more extensive planting), and further inland for sugar cane, Liberian and common coffee, and cacoa. On getting under weigh we tacked out of the river, but the wind being against us we landed again in the evening. Next morning we endeavoured to beat through the pass in the reefs, but the tide had turned, and the wind being still contrary, a furious sea was raised in the pass, so that we were obliged to proceed inside the reefs, rounding Kobalau point, and reaching Savu-savu in the evening. The gentleman to whose kindness I have just alluded took me to his home, and a gale setting in soon after, I was detained there for a few days. On leaving my kind entertainer, I travelled along the south coast of the island to Vatu Kali, near "Fawn Harbour," arrangements having been made for the cutter to follow. There is quite a number of settlers on this part of the coast, who have fine plantations of cocoa-nuts and thriving herds of cattle. To several of them I was indebted for hospitable entertainment. All this coast is well adapted for growing the cocoa-nut tree, which, although now abundant, could be vastly increased. The soil is invariably good, and in places on the sides of the low hills and on the flat lands sugar cane would thrive admirably, as also the two kinds of coffee already so frequently mentioned.

The rocks noticed were agglomerate, basalt, and coral, the first and last particularly common. We left Vatu Kali in the cutter at daylight, but, owing to calms and light winds, and these contrary, it was evening before we arrived at Wairiki in Taviuni. There I was soon at home with old friends and acquaintances, who accompanied me in my journeys through Taviuni.

This is a very fertile island, and capable of producing large quantities of sugar, coffee, and cocoa-nuts. There are considerable numbers of the latter, but there is room for twice as many. I visited several coffee plantations which have lately been made, especially at Gali, and at Messrs. Smith and Aitchinson's, and Forest Creek. These give great promise of success. At na Sali Levu a large sugar estate is in full operation, and the mill is capable of making from two to three tons of sugar per day. The sugar cane grows to perfection here, and in the island there is room for more than thirty such mills. These would be also well supplied with canes by the numerous settlers, who for want of mills are now debarred from turning their attention to this cultivation, and are at present at a loss to know what to do with their land. Some of them grow maize, cocoa-nuts, or cotton, and a few rear stock; but all would prefer growing canes, for which both soil and climate are well adapted, and which would pay better than any other article of production. The area of the island is computed at about 217 square miles, of which about 45 near the coast could be planted with cocoa-nuts, 45 with sugar cane, and 45 with coffee, leaving 82 square miles for forest reserves and waste land.

When passing through the village of Somosomo Taviuni, we called on Tui Cakau, Roko of Cakaudrove province. He was absent; had in fact gone with the men of the village to plant their yams, and to superintend and assist with the work.

On seeing a *meke*, a native dance or play, accompanied by songs, I was surprised to observe the exact manner in which time was kept by the numerous actors. It is the result of practice and many patient rehearsals from youth upwards. In walking through this village we came on a party of about 30 people, who, in preparing for a festival, were practising and reciting the piece which they were to perform for the amusement of their friends. An old man, lying on his side on the grass, but in full view of the actors, was watching the rehearsal, criticising and correcting the faults of the actors.

The province of Cakaudrove includes Taviuni, extends in Vanua Lavu to Savu-savu bay, and includes all the land surrounding Natawa bay. The Roko keeps a smart schooner yacht of about 70 tons, and a cutter, also manned by Fijians, to carry him to different parts of the province, or of the group, either on business or pleasure. The Fijians are owners of numerous rowing and sailing boats, and small vessels of all sizes, up to that of the schooner just mentioned. These craft have, in a great measure, taken the place of the large sea canoes, for which these islands were once noted. The Fijians are hardy if not careless sailors of boats, and, directly or indirectly, they are good customers to the excellent boat builders of the group. These, to some extent, have taken the place of the native canoe builders.

The Fijians have quite a number of athletic games, among which may be mentioned, throwing the *tinika* or reed, wrestling, and one not unlike skittles in some respects. In this game, pillars about a foot in height are built with small stones, and in such a manner that they fall when slightly touched. Several of these pillars (the number varies from 3 to 12) are built near each other in a row, or in half a circle. The player stands at about 30 yards from the pillars, but with his back or side to them, and tries, by throwing a stone from behind, to knock the pillars down. Two others place themselves between the pillars and the player, one on each side, but a few yards from the direct line between the player and the pillars, and with stones endeavour to strike the stone thrown by the player on its passage from him to the pillars. This they frequently succeed in doing.

Throwing the *tinika*, or reed, is practised in the village *Rara* (square), or on a piece of ground smoothed for the purpose. The *tinika* is an oval-shaped piece of heavy wood, about 4 inches long and 2 inches in diameter at the thickest part. A reed or cane about 3 feet long is inserted at the small end, and the game is who will throw or shoot it farthest. The thrower, balancing the reed as in throwing a spear,—his arm stretched at full length, steps backward a few paces, then rushing forward gives the throw the weight of his body as well as his strength of arm, and shoots the reed straight out with great force to a distance of over 300 yards, including the distance which the reed skims along the ground.

Returning to Levuka the steamer called at the island of Mokogai, on which cocoa-nuts are well represented, and nearly all parts of which could be turned into groves of this useful and valuable tree. Copra is extensively made there, and there is also "plant" for extracting fibre from the husk of the nuts. Sheep are bred in the island, and several good flocks of them were seen grazing on the hills.

With respect to communication with the outside world, Fiji is not badly off. Twenty-four hours after the arrival of the mail, viâ San Francisco, at Sydney, a fine steamer of 1,000 to 1,500 tons, belonging to the Australasian Steam Navigation Company, leaves for Levuka. The voyage occupies seven or eight days. The steamer remains at Levuka nearly a week, and leaves with the mails for England in time for them to be transhipped to one of the Peninsula and Oriental steamers at Sydney. The mail service is once every four weeks; and for its punctual performance the Australasian Steam Navigation Company of Sydney is under contract with the government of Fiji. By the same contract the company has to maintain the steamer so often mentioned in the preceding pages to trade in the group, to carry the mails to the different islands, to take cargo to and from them, or for shipment on board the mail steamer. From Melbourne there is direct steam communication to Suva and Levuka about once every five weeks. The owners of this steamer have also a smaller one which collects cargo at different places in Viti Levu. From Auckland in New Zealand several sailing vessels, schooners for the most part, arrive at frequent intervals. (Since the above was in print regular steam communication has been established between Levuka and Auckland, and also between Levuka and Tonga—Friendly Islands.)

CHAPTER II.

FLORA.

The Flora indigenous to Fiji, as far as known, amounts to 1,086 species of flowering plants, and 245 species of ferns and allied plants. Of these numbers 635 species (620 flowering plants and 15 ferns) have as yet been met with only in Fiji. This number (635) of endemic species, out of a total of 1,331 species of ferns and flowering plants, is very large, and seems strikingly peculiar. But this is, in a great measure, owing to the fact that of all the Polynesian islands, the Fiji group is the one which is botanically best known. When botanists are better acquainted than they are at present, with the Flora of the interior of the Samoan, New Hebrides, and other Polynesian islands, it will be found that many of the species, which are now considered Fijian, have a wide distribution throughout Polynesia. Notwithstanding all that has been done by the United States Exploring Expedition, Milne, Dr. Harvey, Dr. Seemann, and myself, to make the flora of Fiji well known to Science, much yet remains to be done, and I regard the discovery of new plants in the group as far from being exhausted.

To the number of indigenous flowering plants described in Seemann's Flora Vitiensis, which included all the plants that had been discovered in Fiji up to the date of its publication, 3 natural orders, 34 genera, and 363 species were added by my visit. Of ferns and allied plants, I discovered 15 new species, beside finding in Fiji 20 other species, which had not previously been found in Polynesia. Thanks to Mr. Baker, of the Royal Herbarium, Kew, the ferns I collected in Fiji were most carefully and critically examined.

The flowering plants have not yet been so carefully compared. Therefore, I anticipate that when these 363 species of flowering plants, which I at present regard as new, are minutely examined and compared with old and well known species, the number will be reduced about 20 per cent., or to 300 species. Thus, one of the results of my visit will be that I have discovered, or added to the known Flora of Fiji 300 species of flowering plants and 35 ferns.

The largest orders are Leguminosæ, which is represented, in the group, by 36 genera and 62 species; Rubiaceæ by 23 genera and 122 species; Orchids by 25 genera and 49 species; Euphorbiæ, and Urticaceæ 20 genera each, and species 79 and 52 respectively. About 130 species are common to Australia, but the greater number of these are found in many parts of the old and new worlds. Many American plants have found their way to Fiji through, it is presumed, the islands lying between that continent and Fiji.

Several weeds, common in other countries, have also found their way there, through the agency of civilized man. Some of these are now his pests, destroying his pasturage and giving a great amount of trouble and expense in keeping his plantations clean. Most of the sea-shore plants found in Fiji, have, as might be supposed, a wide distribution. Several species of what are known as sea-shore plants are also found far in the interior of the largest islands of the group. The most noticeable were cerbera odallam, afzelia bijuga *vesi*, heritiera littoralis, calophyllum inophyllum *dilo*, ippomoea pes-capræ, kleinhovia hospita, pandanus odoratissimus, cynometra grandiflora, cassytha filiformis, cocoa-nut, terminalia catappa *tavola*, paritium tiliaceum *vaudina*. The presence of the cocoa-nut and cerbera in the interior as well as the *dilo* and *tavola* may be owing to the agency of the natives. On the other hand, several species of land plants are found on the sea-shore, on the edges of tidal estuaries, and salt water marshes. The branches of large trees overhanging the sea may frequently be found covered with epiphytes, or orchids, ferns and lycopods, which all apparently rejoice in an occasional bath of salt water spray. None of the mountains of Fiji are high enough for an alpine flora to exist. Many of the plants found on the tops of the mountains are also found near the level of the sea, even on the shore. On the other hand, sea level plants may also be found on the tops of the hills. But in the one case, the number of sea-level plants decreases in the ascent to a higher elevation, and in the other, mountain plants become fewer on approaching the sea. Except in favoured situations, plants whose habitat is in the warm sheltered plains and valleys, present a stunted and weather-beaten appearance, when met with on the higher mountain sides. This in general may be owing to the scantiness of the soil as well as exposure to the winds and cool air. The plants of the mountain tops, when found near the sea still occupy a position similar to their native one, viz., the top of a ridge. But when these plants are found at low elevations, the majority of them are lank, drawn up, and lean on others for support, plainly indicating that the climate is too hot for them.

One or two species of cinnamon, litsea, alstonia, paphia, polystichum, blechnum, calophyllum, podocarpus, calanthe, pavetta, selaginella, and a great variety of mosses are most commonly found on the top of the mountains. Astelia, calophyllum burmannii cinnamomum pedatinervum, gnetum genom, podocarpus cupressina *kau labua*, podocarpus viticnsis *dakua-salusalu*, dacrydium elatum *leweninini*, dammara viticnsis *dakua*, kentia exorrhiza *niu soria* or *niu sau*, and some other palms may be found thriving on the sides of the mountains, from the sea-shore up to elevations of 2,500 or 3,000 feet, *i.e.*, when in situations sheltered from prevailing winds and in soil favourable for their development.

The windward side of the islands, where most rain falls, is one dense tangled mass of vegetation, through which it is impossible to force a passage without great exertion. This is especially so on land that has at one time been cultivated, but abandoned for some years. Land thus abandoned is at once overgrown by reeds *gasau*, wild sugar canes *vicos*, sponia orientalis, and sponia velutina, homalanthus populifolia, tree ferns, large growing climbers, &c., all contending for the mastery. This goes on until perhaps the growth is checked by fire, which burns up the grass, the leaves of the sugar cane, &c. The hardier kinds of exogens that have been least injured by the fire, start into new and vigorous growth and ultimately take possession and give shelter to hardy ferns, zingiber zerumbet, and climbing freycinetias. The tender ferns follow, accompanied by shade loving alpinias, heliconias, pipers, palms, cyrtandras, &c. Along with these, young plants of podocarps, conifers, calophyllums, &c., make their appearance. After a few years,—for growth is rapid in the moist warm climate of Fiji,—the woods begin to assume the appearance of a virgin forest, with a dense undergrowth of ferns and shrubs.

The branches of the large trees are covered with orchids, lycopods, ferns, and loranths, while their gigantic trunks are covered with hymenophyllums, and trichomanes, climbing pipers which give them the appearance of columns clad with ivy. Such epiphytes as pothos scandens, and other orontiads make the trunks of the trees look like pillars clad with creeping caladiums; branching freycinetias, &c., make the trees to which they cling appear like tall columns of grass, the tops of which are loose and wave backwards and forwards in the breeze. Sometimes the trunks may be seen all aglow with the bright flowers of a creeping medinilla, adorning as with wreaths the tree to which it clings, and adding beauty and variety to the agreeable though somewhat sombre shade of the forest below, while its leaves are seen mingling with those of the tree, far above. A break in the forest displays festoons of white flowering clerodendrons hanging from the top of a tall tree, while a small gap will be covered over like a trellis with the beautiful carruthersia and sweet smelling jasminiums.

Looking down upon the forest, from an eminence, it presents one mass of dark green foliage, the uniformity of which is broken by the delicate fronds of the tree fern, and the waving leaves of the palms, and varied by the gay flowers of ipomœas and other climbers which have overtopped the trees and are spreading themselves out to the light and air.

The leeward side of the largest islands is an open country,—undulating hills covered with grass and ferns, with patches of forest in ravines and tops of mountains,—places which have escaped both the ravages of fire and the migratory cultivation of the

Fijians. Such patches are comparatively few. In them the hard-leaved myrtaceæ abound. Numbers of *noko-noko* casuarina equisetifolia; the *gumas*, acacia richii; geitonoplesiums and a species of hibbertia, along with the open grassy aspect of this part of the country, give it a striking resemblance to some parts of Australia. The dark foliage of the *kau kuru*, casuarina nodiflora, in dense masses, has the sombre aspect of a pine-clad mountain side. While the presence of screw pines and sago palms (cycas circinalis) in clumps or single specimens dotted over the country, give the landscape a somewhat antediluvian aspect. Acres of land are covered with ferns, the "bracken," pteris esculenta, gleichenia oceanica, and nephrodium molle.

The "turmeric" plant (curcuma longa, *cogo* of the natives) abounds in this part of Fiji, and hillsides may be seen covered with it, mingled with the *yaka* (pachyrrhizus angulatus), and the *yabia*, Fijian arrowroot, several varieties of tacca. In most of the dry districts, several species of the gardenia abound, their white flowers adorning the landscape and perfuming the air with their odour, which, in the open atmosphere, is refreshing and pleasant. The "candle nut" (aleurites triloba, *lauci*), is the most common tree in the ravines. The hoary grey colour of its young leaves, contrasted with the dark green of the older ones, and the adjacent tree ferns, gives a pleasing variety to the whole. These localities are the habitat of the scented woods of Fiji, the sandalwood, the *bua-bua* and the *savoo*. On rocky clefts the odoriferous and pretty bua, fagraea berteriana, and several species of drymispermums abound, the last in company with blue, white, and red flowering cranthemums.

The sides of streams in these dry districts are shaded by such trees as heritiera litoralis, afzelia bijuga, terminalia catappa, kleinhovia hospita, cynometra grandiflora, *cibicibi* (pr. thimbi-thimbi), bamboos, calophyllums, eugenias, &c. Away from the immediate vicinity of the streams these trees are deciduous, calophyllums and eugenias excepted.

That pretty and fragrant white flowering genus of small trees, dolicholobium, is found from the sea-shores to the tops of the highest mountains in both wet and dry districts. Numerous specimens of gay flowering hibiscus of several species also abound. Along with these red leaved and green leaved dracaenas and crotons, with their leaves spotted yellow, green, and red may all be seen. Shady woods, in both dry and wet localities, are bright with the flowers of several sorts of small shrubs (ophiorrhiza), and the larger, but not less pretty flowers of many species of psychotria.

The rocky banks of many streams are adorned with the cream coloured flowers of lindenia vitiensis, impregnating the air with their sweet odour. The slender, half-climbing shrub, mussaenda frondosa, with its golden flowers, large white phylloid calyx and green leaves, decorates many an acre of waste grassy land, where the orange coloured dove and the red and the green parrots flit to and fro.

Ferns abound everywhere,—from sea level to the highest mountain tops,—in the hottest and coldest parts,—in sunshine and shade,—on the poorest and richest soils, and in driest and wettest parts. They are of all sizes, from the tiny hymenophyllum, scarcely one fourth of an inch, to the gigantic alsophila, tree-fern, having a trunk 50 feet or more in height, surmounted by a crown of beautiful feathery-looking fronds. The number of distinct species and varieties of ferns and allied lycopodiums and selaginellas indigenous to Fiji amounts, as yet found, to 246. I have no doubt the number will be raised to about 300 species when every part of the islands has been explored. Some of these ferns are magnificent. The dicksonia moluccana has fronds of a triangular shape, measuring 12 feet in length and 10 feet in breadth at the base. One of them would cover an area of 60 superficial feet. This gigantic leaf is supported by a stipe or stalk 6 feet in length and 3 inches in circumference. As a contrast to this may be mentioned the tiny fronds of the flimy ferns, hymenophyllums, and some species of trichomanes, scarcely an eighth of an inch in size. The delicate fronds of a few species of the last named genus attain a height of $2\frac{1}{2}$ feet. Most beautiful they look when seen with the rain drops hanging like beads of crystal from the points of their finely-divided fronds. Not less pretty in this respect are hymenophyllums javanicum, and dilatum, generally found on the sides of streams, shaded from the sun by the overhanging banks and lofty trees.

The davallias found in Fiji are worthy of notice. However, the most beautiful of them are hymenophylloides, and blumeana. The latter is, without doubt, the prettiest fern in Fiji. Both are found on the loamy banks of streamlets in densly shaded woods. Their fronds seldom exceed 1 foot in height, are of a pale green colour, finely divided, and their membranous texture gives them a pellucid crystalline appearance, especially when they are covered with dew. In the dry parts of Fiji, one of the silver-leaved ferns (cheilanthes farinosa) may occasionally be found growing in the crevices of the rocks, and its pretty relation, cheilanthes tenuifolia, which, with pteris ensiformis, and pteris geraniifolia, abounds in dry grassy fields, and comes up after the rains. While festoons of lygodium reticulatum *wa kalou* (holy creeper), and tassels of lycopodiums phlegmaria, and nummularifolium, to 5 feet in length, hang from almost every tree, the surface of the ground below is clad with one dense mass of beautiful selaginellas, some of which attain a height of 4 or 5 feet.

There are not many different species of orchids in Fiji, but the various members of several genera are well represented, and are not unworthy of notice. A species of calanthe, having a spike of white flowers spotted with red, which grows among grass, is common and beautiful. Another species, of the same genus, with snow white flowers, abounds in shady forests in wet and dry localities; and a third, having beautiful orange coloured flowers, sessile and clustered together on a short spike like the flowers of a hyacinth, was found in one place near the top of Voma Peak.

Several species of dendrobium are worthy of cultivation; these are mohlianum, tokai, Gordoni (n. sp. Le M. Moore), and Hornei (n. sp. Le M. Moore). The latter was found in the island of Rabi, growing on a tree on the sea-shore, where it was occasionally bathed in the salt spray of the breaking waves. Dendrobium Gordoni was found in Samoa, growing on a pandanus tree in a swamp, in the island of Upolu, and also on an old *dakua* tree (dammara vitiensis) near the Blackwater, (*Wai ni Loa,*) in the interior of Viti Levu.

Among parasites there are three species of loranths, and two curious species of apphylous viscums (misletoe). The loranths are very common in Fiji, and are found on nearly every *ivi* tree (inocarpus edulis). The flowers of the loranths are beautifully coloured, red and purple, yellow, and yellow and purple. The hydnophytum is a curious and interesting genus of rubiaceae. Four species of this genus have been discovered in Fiji. They are epiphytal plants, and are generally found on the trunks or branches of trees. Their favourite position is between the forks of the branches, where they sit most securely. The part that may be termed the stem or trunk of the plant, is cone shaped, flat or concave at the base, adapting itself to that part of the trunk or branch on which it sits, and rounded more or less at the top or apex. The largest seen was about a foot in diameter, and 15 inches in height. The stem is composed of a soft, spongy, fibrous substance, dark coloured on the outside, altogether, not unlike a large purple turnip. The black fibrous roots are generally emitted from the edges of the stem at its base, and cover, like a net, the bark of the tree to which they firmly adhere. From the top of the stem rise the branches bearing the leaves and flowers. The latter are either white or yellow, and sometimes an inch in length. The stem is a favourite residence of vicious black ants which make their nests in it, and hollow it out into numerous passages, from which they issue in hordes at the least disturbance. Balanophora is a curious root parasite, and the *waloa,* or black creeper (rhizomorpha), is an interesting fungus to the European, and a useful one to the Fijian.

There are comparatively few genera of palms indigenous to Fiji, but the several species are numerously found and well represented. The *niu sau* (areca or kentia exorrhiza) is a tall graceful palm, frequently rising to a height of 80 feet in favourable situations. It is most common in low-lying districts, but may be found on the tops of the highest mountains. Pritchardia pacifica, the *niu masie* of the Fijians, and the fan palm of the settlers, although not rare, cannot be said to abound in a natural or wild state. It is a handsome, and in some respects a useful species. Some species of ptychospermum, the *cage-cake* (pr. thangethake), and the *balaka*, are very handsome trees. One or two species of the genus attain the dimensions of the *niu sau*, while others do not exceed the size of a stout cane. These latter are found in the dense virgin forests of the wettest parts of Fiji, and are common from sea level to the mountain tops. A kind of sago palm (sagus vitiensis), *sogo* is not uncommon in some of these islands. It is most abundant on low-lying swampy land, on the banks of the Navua and the Wai Manu rivers in Viti Levu, and occasionally met with in other islands. It attains a height of about 35 feet, and is a strong growing picturesque tree. The inflorescence is a large terminal panicle, 10 feet or so in height, and about 7 feet in diameter at the base. After flowering and bearing seed the tree dies.

Although the leeward side of the large islands is an open grass country, yet there are comparatively few kinds of grasses indigenous to Fiji. However, the species that exist are well represented. The paucity of species may be owing to the fires which annually sweep over the dry districts, burning up all the herbage that springs up during the wet season. The *gasau* or reed (eulalia japonica), is by far the most abundant, covering whole sides of hills and sometimes descending into the flat low lying lands. An unnamed species, the *drauka*, although indigenous and not rare, is generally found cultivated. Sugar canes *dovo* (saccharum officinarum), are common; both wild and cultivated varieties. The wild varieties grow in dense brakes on the rich alluvial flats and along the sides of small rivers and streams. They frequently grow to a height or length of about 20 feet, with a diameter varying from one-fourth of an inch to an inch. They are of various colours, green, white, or red, and some varieties are striped like a ribbon. The juice of some of the varieties has a faint sweet taste, but that of the majority is insipid and watery. Their characters at once suggest them to be the plants from which the cultivated varieties of the sugar cane have descended by improvement on successive sorts from a distant period. Improvement on them will be tried in the Botanical Gardens at Mauritius. Such experiments while interesting to botanists will be of great importance to growers of sugar cane. Bamboos, of which there are several kinds, are numerous. Lemon scented grass *caboi*, [cymbopogon refractus, and perhaps another species] is abundant in many places of some of the islands.

There are very few species of sedge in Fiji, but like the grasses the various species are plentifully represented. They are generally confined to swamps over which fire sweeps annually in the dry season, destroying all the vegetation dry enough to burn, and exterminating the annual and the tender kinds.

In arboreous vegetation, Fiji has many fine specimens. Although none of the trees attain the dimensions of the mammoth trees of California, nor the gigantic size of some Australian eucalypti, yet they are splendid examples of their respective kinds. On the Tai Levu coast a tree of barringtonia speciosa, *vutu*, when measured at 6 feet above the ground, was found to be 33 feet in circumference of trunk. The trunk was about 12 feet in length, the loftiest branches were not more than 40 feet from the ground, but so wide spreading that, although the points of the branches had been lopped off, they still covered an area of about half an acre. The charred trunk (half of the trunk was burned) of a *dilo* tree (calophyllum inophyllum) growing on the shore at Rabi measured 7 feet in diameter, by a length of 12 feet. The head of the tree was proportionately large, and overhanging the sea. Some magnificent trees of *cibi-cibi* (cynometra sp.), *vesi* (afzelia bijuga), and *vaivai* (serianthes myriadenia), were seen in the mountains of the same islands. The trunks of a few of the first named would give an average diameter of 3 feet on a length of 40 feet; those of the second 2½ feet on a length of 30 feet; and the third 2½ feet on a length of 20 feet. Equally good samples of these trees may be seen in other parts of Fiji. On the banks of the Tamavua river, in Viti Levu, there is a tree of *dakua* (dammara vitiensis) nearly 100 feet in height. Its trunk, when measured at 6 feet above the ground, was found to be 25 feet in circumference. At about 20 feet from the ground the trunk had been broken, and is now divided into a number of upright growing shoots, each of which has the dimensions of a tree of more than medium size. In the mountains of Taviuni are many fine specimens of the *damanu* (calophyllum burmanni) and *dilo-dilo* (calophyllum spectable, of the U. S. exploring expedition), as well as other kinds whose dimensions would equal either of the *cibicibi, vaivai*, and *vesi* at Rabi. Trees of that size are by no means rare in the virgin forest of Viti Levu and Vanua Levu, and the *dakua* (dammara vitiensis), *dakua salu-salu* (podocarpus vitiensis), and *lewininini* (dacrydium alatum), may be added to the list of large trees common in these two islands, but not in any of the other islands of the group. In the forests which lie between the wet and dry districts, and also in the mountains in the wetter parts of the latter, trees of the largest size are most numerous. These forests are composed of prime timber trees, such as *dakua, dakua salu-salu, lewininini, damanu, kau tabua*, and several other sorts, whose trunks will give an average diameter of from $1\frac{1}{2}$ to $3\frac{1}{2}$ feet on a length of about 40 feet.

The *ivi* or "Polynesian chestnut" (inocarpus edulis) is an interesting tree. It is rather larger than an average sized English elm, frequently growing to a height of about 80 feet, and in the outline of its head and habit it is not unlike that tree. Its trunk is a curiosity. It is deeply fluted or rather buttressed all round, and a section of it would not be unlike a cart wheel, minus the felloes; the buttresses, like the spokes, springing from a central part resembling the nave. The diameter between the extremities will range up to 20 feet, that of the central part (nave) perhaps a foot. The *dawa* (nephelium pinnatum) is another noble tree found in all parts of Fiji. It is deciduous, and when the young leaves unfold themselves, their peculiar reddish green colour gives a pleasing variety to the landscape. Before the leaves fall off they change from green to a reddish brick colour.

Several kinds of large fig trees *baka* (ficus) are found in Fiji, notably one at Bureta in Ovalau, another on the Rewa in the interior of Viti Levu, and a third between the native towns of Vose Dam and Wai Basaga on the Siga Toka river. The last-mentioned tree serves as a sort of half-way house between the two towns. Its trunk is hollow, and could shelter 20 or 30 men. Wayfarers kindle fires in it to cook their food and keep themselves warm during the night. These large fig trees commence life as epiphytes, and ultimately strangle the tree which supported them. A seed has been dropped, perhaps by a bird, on a branch in the fork of the tree; the position being favourable, the seed germinates, and a tender radical, like a thread, grows down the side of the trunk, clinging to the rough bark for support, till it reaches the ground. At the same time a stem grows upwards from the seed. This root increases yearly in size and strength, so does the young plant, which sends, annually, instalments of roots to the ground, in the same way as the first. The thread-like roots rapidly grow thicker after they enter the ground, and resemble perpendicular columns all round the trunk of the tree. Another set of roots wind round these, like a many-folded net. These roots do not displace one another. They unite perfectly, as by affinity, whenever they come in contact, ever tightening their hold on the trunk of the tree, which, at last succumbs under the strain, decays, and its place is left hollow. Some other kinds increase by dropping rootlets from their branches, and spread until they cover large areas of land, and it is impossible to say, with certainty, which is the original from whence all have sprung.

The flora of Fiji is essentially tropical. A few species belonging to a temperate one may however be met with on the mountain tops. Its general character is Polynesian, with some affinities to the flora of Australia and the Malay Islands on the western and north-western sides of the group. These resemblances decrease towards America on the east, and the northern parts of New Zealand on the south.

An alphabetical list of the plants will be found in the appendix, and also a list of the genera under the order to which each belongs.

CHAPTER III.

Fijian Food Plants—Method of Cultivation—Vegetables.—&c.

The instincts of the Fijian are agricultural, and it may be said that he finds a use for all the vegetable products of his country, and has a name (sometimes several) for each individual plant. His knowledge of their use is seldom at fault, and they provide him a never failing supply for all his needs. Of food plants he cultivates yams, *dalo* or *taro*, sugar cane, *li* dracæna sp., bread fruit, &c. With great aptitude he selects from the forest portions of land best adapted for the several kinds of crops. The effects of nature on the vegetation which surrounds him are his guides to the season for digging, planting, and gathering in his crops. Of cereals he cultivates very little, maize for Government taxes being the chief.

The people live principally on yams, dalo, bananas, bread fruit, with fish, fowls, pork, and several kinds of greens. Their drink is generally water, or the milk of the cocoa-nut. When tired, or on festive occasions, they use *agona* or *kava*. The yam, *uvi*, is their staple food; they have about 20 different sorts under cultivation. Some of the varieties are very fine, mealy, and free from fibre, like a good potato. The tubers of some of the kinds do not exceed two or three lbs. in weight, but those of one or two sorts weigh as much as 100 lbs. in weight. The Fijians say that the yam thrives best in hard or rough, unprepared ground. This may very probably be owing to the ground being new and rich.

The trees having been felled, and their roots cleared out of the ground, and the grass, &c., burnt off, planting commences when the *drala* tree (erythrina indica) begins to flower, in the month of July or August, according as the season is late or early. All hands assist in planting. The soil is thrown up into small mounds, *buke*, from 3 to 5 feet apart. On each mound a small yam, or the crown of a large one is planted. Should the ground be flat, open drains are made to carry off the water, or the ground is thrown up into beds with ditches between, and the yam mounds formed in rows on the beds. The stems of the yams are supplied with canes to climb upon. The canes are generally laid horizontally, supported by forked sticks stuck in the ground, or by the tops of the mounds. The roots or tubers are ready for digging by the beginning of March; the drying of the stems indicates that the tubers are ripe. When dug, the yams are stored in airy sheds erected on the fields. These sheds are constructed of bamboos set upright, and the roof made waterproof by a thatching of grass. After being stored, the yams are turned over occasionally; the young stems are rubbed off those that have started growth, and all the decaying ones are removed. They are used boiled, roasted, or steamed, the larger ones being cut in pieces, and the smaller ones cooked whole.

There are two kinds of *dalo*, land and water *dalo*, just as in India there is mountain and swamp rice. The land *dalo* is most commonly met with in the wettest districts. In fact, like mountain rice, it will only grow in places where the rainfall is great, and on land from which the water has neither drained, nor evaporated after the forests have been cut down.

To prevent too rapid evaporation, a few trees are left growing on the land selected for the *dalo*. To prepare the ground for *dalo*, the trees are felled and burned, and the ground cleared from roots of grass weeds, &c. The plants are put down in rows from 2 to 3 feet apart, and less is allowed from plant to plant. Holes about 9 inches in depth are made in the ground with a planting stick, which, before being pulled out, is well shaken to harden the sides of the hole, prevent the ground falling in, and water from passing freely through the soil. This hole catches a considerable amount of rain water, as it runs off the ground, and generally retains it during a protracted drought. Besides keeping the *dalo* moist, the depth at which it is planted prevents it throwing out suckers at the top of the tubers (corms), which it has a tendency to do when planted near the surface. The part planted is a thin piece cut from the crown of the tuber, with the leaf stalks attached,—the leaves being removed to prevent exhaustion. The Fijians seem to be aware of this, and in transplanting the *masi* (broussonetia papyrifera), bread fruit, &c., they remove the leaves when lifting the plant. The only care bestowed on the land *dalo* after planting is to keep down weeds, and this weeding is generally done in wet weather, lest the ground should be too suddenly exposed to the drying influences of sun and air.

The place selected for growing water *dalo* is generally the bottom of a valley, or any place where water is at command. This plant, like rice, requires a constant supply of fresh water. Not a little labour and ingenuity has been displayed by the Fijians in making aqueducts, often miles in length, over ravine and hollow, to carry a supply of fresh water to these plantations. Sometimes the hill sides are terraced for it, as is done for paddy or rice in some parts of Ceylon. Of course, they are unacquainted with the use of the spirit-level, and the level must have been obtained by digging and allowing the water to follow. This would indicate the highest point to which the water would rise. In making the retaining walls for the beds and terraces on the hill sides, a good deal of labour has been bestowed. One bed follows another in succession, the fall from a bed to the one next below it varying from 4 inches to several feet, according to the steepness of the site. The beds also vary greatly in size. The settlers call them "*dalo* or *taro* patches." As in the case of yams and land *dalo*, these beds, after a crop or two has been taken off them, are allowed to lie fallow for several years, in order that the land may regain its fertility.

When the land is to be re-cultivated, the grass, &c. is cleared off the surface, and either burned or thrown aside; the aqueducts and retaining walls are put in order. The ground is dug and the surface levelled,—water being let on frequently for a trial of the level. The *dalo* is then planted in the same manner as the land *dalo* above described, only that the plants are generally put down at a less distance apart. The *dalo* grows best in a heavy stiff clay soil, and generally takes from 10 to 12 months to reach maturity. The decaying of the leaves indicates that the tubers are ripe. The tubers vary in weight from 1 to 12 lbs., if they have been well grown. The average weight, however, is from 4 to 6 lbs.

They are eaten either boiled or roasted, and are starchy and very nourishing. The natives prefer eating them cold rather than hot. The *dalo*, along with some sweetening matter, frequently forms the chief ingredient in the native *vakalolo*, or puddings. The young leaves, boiled and served like spinach, are an excellent vegetable. Like the corms, they are extremely acrid when raw, but the acridity is removed by the heat when cooking.

There are about 18 different varieties of the *dalo* cultivated in Fiji. They differ from each other principally in the size and colour of the leaves and leaf stalks. Some of the varieties are very handsome plants, deserving a place in any collection of plants for hothouse decorations. The natives interchange the tops for planting from one district to another, from a cold district to a warm, from a wet to a dry, and from one kind of soil to another. The Fijians know, by experience, that these changes are beneficial to the *dalo* plant, and that a larger and better crop is obtained than by constantly cultivating the same sort in the same district or kind of soil. When the "tops" are ready for planting, before the ground is in proper order to receive them, they are preserved by partly immersing them in water, generally in some place where they are shaded from the sun.

Although the Fijians are well aware of the benefit derived from manure when applied to their "pet" sugar canes, they do not in general make use of it for any of their crops. In selecting a piece of land for growing their yams, &c., they are generally correct as to the kind of land best adapted to the purpose, but as to the choice of the *situation* they seem to be moved by fancy. They have a large area to chose from, and the spot selected may be as far as 5 or 6 miles from the town where they live. Not because there is no land suitable for the purpose nearer, but apparently merely to gratify a whim, or fancy. Their time is mostly at their own disposal. A kind of temporary dwelling is erected close to the selected land to which they remove with their families during the season for clearing the land and planting the crop. After this is completed, they return to the town (Koro), and occasional visits are made to the plantation for the purpose of weeding, &c., and to see how the crops are growing. The common people generally assist the chief in the heavy part of his agricultural labour, clearing the land, &c., and the poorer people are aided by their wives. The latter have to carry home from the fields all the yams, canes, &c., the men evidently thinking such labour beneath their dignity; or, more likely, they are debarred by custom from helping the women.

Their cultivation is migratory. They seldom take a crop of the same piece of land for two successive years, except in well watered districts, when a crop of land *dalo* may be found succeeding one of yams, or the reverse. One crop, however, is the rule. The land is then abandoned for an indefinite period, until its fertility is restored. When such a system of cultivation prevails, a large area of country is required to supply food to a comparatively small population. In the elevated parts of the windward districts, little injury is done by the unwooding of much forest land annually for plantations — an equal extent being annually abandoned. After an abundant fall of rain, so rapid is the growth, that the land thus abandoned is speedily covered by a dense growth of trees. In this way a kind of balance is effected, provided that upon such land, in after years, fire is prevented from entering and destroying the trees which are grown on it. This, unfortunately, occurs too often, as the low-lying and grass-covered hills on the windward side of Ovalau, Vanua Levu, Viti Levu, and other islands testify. This has, also, been the cause of unwooding the leeward parts of Viti Levu, and Vanua Levu, the consequence of which is that these districts are often weeks without a shower, and the land is parched by drought, and rendered almost barren. In short, it is apparent that, with a dense population to support, and the annual requirement of new land whereon to grow food crops, if this system of agriculture be not abolished it will bring ruin on the whole country. In the cultivation of yams and *dalo*, or in the management of a canoe on a rapid river, the Fijian could give a lesson to a more civilised being; but, for the general welfare of the country, it will be necessary to teach him another system of agriculture, by showing him that by the use of manure, crops of the same sort can be taken off the same land for a succession of years. The introduction of cereal food, as rice, flour, corn, &c., for the use of the people, would help greatly to alter the present vicious system of cultivation.

In digging and preparing the soil, forming aqueducts, &c., the Fijians seldom use any other implement than a pointed stick made of some hard and tough wood, and the hands. Even now the spade or hoe is seldom used, but they are becoming more common than formerly; and the use of these and other implements will in time prevail. The native plan of digging is as follows. The men provide themselves with a digging stick each, and by repeated blows of these make holes round a piece of ground of about 2 feet in diameter; then by using the sticks as levers this piece of soil is turned over on its side, or upside down. Boys follow, and break up these lumps by blows from short sticks, pulverizing the soil with their hands.

Among native food plants the banana may be ranked after the yam and *dalo* in order of merit. This is more especially the case in the interior of Viti Levu, where the people have not so many cocoa-nut trees as those who live on the coast, and in the smaller islands, where the cocoa-nut abounds. Even then the bananas are largely used for food, roasted (when green), raw (when ripe), also cooked with cocoa-nut milk and the juice of the sugar cane, as a *vakalolo* or pudding. Banana plantations abound everywhere, and extensively so in *Colo*, *i.e.*, the mountain districts of Viti Levu. They are planted along the sides of the road to shade the traveller from the sun, sometimes forming avenues miles in length or more. The fruit on these trees is *tabu*, that is *forbidden* to travellers. The *tabu* is invariably respected by the natives.

Bananas are planted in rows, and the trees are put down at about 8 feet apart, and the same distance is allowed between each tree. Suckers from the sides of old roots are used as plants, the leaves being cut off before planting for the reason already mentioned. The soil in the place where the young tree is to be planted, is dug in a circle of about 3 feet in diameter, and to the depth of 2 feet, and well manipulated. The young trees bear in about two or three years after planting. These plantations are frequently formed on land that has yielded a crop of yams or *dalo*. The latter is sometimes planted along with bananas, whose leaves, as the natives say, shade it from the sun, or more correctly they shade the ground, and prevent a too rapid evaporation of moisture. Bananas are extensively exported, but the supply is considerably short of the demand. The leaves of the banana are often used as plates for serving food upon, as tablecloths, and also for wrapping material. When used for the latter purpose, the mid-rib is cut close to the leaf, which is passed several times through the flames of a fire to make it tough and pliable. Thus treated, the leaf does not split readily. The leaves of the dalo are also used for the same and similar purposes, for which they are prepared as the banana. The stems of the common banana yield a fibre scarcely inferior to "Manilla hemp" (musa texilis), but it is not extracted. An abundant supply of it could be obtained from such a country as Fiji, where bananas are planted on a large scale as food to the inhabitants and for exportation.

Next to the above-mentioned food plants comes the bread-fruit (artocarpus incisa) which is a most useful as well as a highly ornamental tree. It sometimes attains a height of 50 feet, but the average is from 30 to 40 feet. In general its trunk will measure about 15 feet to the first branches, with a girth of 3 to 4 feet. It is a horizontal branching tree, with a cone-shaped head. The leaves of the young trees are sometimes 2 feet in length, and from 12 to 15 inches in width. Those of the older trees are little more than half that size. They are covered with rough hairs, which makes them disagreeable to the touch. Some of the varieties have leaves deeply lobed, and those of some others are almost entire. The fruit of some of these varieties weighs as much as 9 lbs.; that of others does not exceed 1 or 2 lbs., and 4 or 5 pounds may be reckoned the average weight throughout the group. They are in general cone-shaped, flattened at the base, or speriod. The quality of some of them is excellent, dry and mealy like a potato; that of others is watery and insipid. They are either baked or boiled, and eaten alone, or with pork or fish. Sometimes they are made into puddings, or buried under ground, and made into a *mandrai, i.e.*, native bread. At all periods of the year there are some of the varieties in fruit, but the fruit is most abundant from the middle of February to the middle of April. In some of the native towns the trees are abundant, and groups of 20, or more, may frequently be seen scattered over land which had been cultivated. Large numbers of the trees were destroyed in the wars that constantly occurred between different tribes,—the first acts of an invading force being to destroy the food plants and fruit-bearing trees of the tribe invaded.

One or more of the varieties of the bread-fruit bear seeds, but the most of them are barren. It is doubtful whether these seed-bearing trees are varieties of the artocarpus incisa, or if they do not form another species of the same genus. The wood of the bread-fruit is used for some purposes by the Fijians, but it is not so good as that of the "Jack" (artocarpus integrifolia) or the artocarpus hirsuta. It is soft, light brown, with parallel veins of a reddish colour. When wounded, the tree yields a large quantity of white sticky juice, which is used for caulking the seams of canoes. The tree is propagated by suckers attached to a portion of the root from which the sucker has sprung. The young trees grow rapidly, and in the third or fourth year after planting they reach a height of about 16 feet, and begin to bear fruit. They have a picturesque appearance peculiar to themselves of which a minute description would convey a very indifferent idea.

The sweet potato, *kumara* (ipomæa batatas or batatas edulis), is cultivated to a small extent by the Fijians, and largely by the settlers, as food for their labourers. There are two varieties, one of which has tubers of a reddish colour, those of the other are white. Both of these are excellent, and it is surprising that they are not a favourite article of food with the natives. These and yams were selling in Levuka last year at from £4 to £6 per ton, according to quality, and 5 or 6 tons can be produced from an acre of land.

Besides the above-mentioned plants which are cultivated for food, there are others of less importance, either growing wild in the forests or found in a semi-wild state in many places. One of these is the *via mila* (alocasia indica). It is found growing by the sides of streams, at the edges of dalo patches and marshy places throughout Fiji. It attains a height of about 10 feet when full grown, and has a handsome, striking appearance. The stem, the part eaten, is frequently about 4 to 6 inches in diameter. It is, if anything, more acrid than the *dalo*, and is only eaten in times of scarcity, or, for a change. In well developed plants, the leaves are about 3 feet long, and from 2 to $2\frac{1}{2}$ broad at the base, smooth, and of a dark green colour. In Seychelles, where the plant is common, and by a curious coincidence called "Via," the large, fleshy succulent stems, or immense elongated corms, are used, when boiled, for feeding pigs. The *via kau*, or *via kana* (cyrtosperma edulis) is sometimes cultivated in swampy places like *dalo*. Although in general use there is not much of it eaten; neither is it so good or so highly esteemed as the latter plant. Both the *via mila* and the *via kana* are eaten either boiled or roasted. Sometimes they are grated, and along with other ingredients made into *madrai*. The poisonous qualities are driven off by the heat in cooking.

The *daiga* (amorphophallus campanulatus) grows wild in almost every part of Fiji. The root, the part used for food, is a flat tuber about 6 inches broad and 3 or 4 inches thick. The plant is herbaceous, and the flower appears before the leaf; it has only one of each. The flower, on well developed plants, is about 6 inches in diameter, and rises about 9 inches above the ground. It has a curious and grotesque appearance. The spathe, or outer floral covering, is brown, and is supported in the centre by the spadix, around which the spathe, when undeveloped, is closely wrapped, —much in the same way as an unexpanded bell-shaped tent is wrapped around the centre pole. As the flower attains maturity the spathe expands like an umbrella, but the margin is rolled inwards. Thus strengthened, the margin is prevented from falling and covering the spadix which, by the expansion of the spathe, is exposed like the handle of an opened umbrella. The individual flowers are sessile, and arranged closely together round the spadix. When the flowers reach maturity, they emit an offensive carrion-like odour, which may be smelt at some distance. Though hidden among long grass, the position of a plant of the *daiga*, when in flower, is quickly ascertained by the offensive smell and the swarms of flies that hover about it. As the individual flowers reach a mature stage one after another the disagreeable smell is kept up for a week or ten days. The leaf, which is handsome and pinnatifidly divided, pedate-shaped, is from

9 inches to a foot in length, and about 18 inches in breadth, of a pale green colour, blotched with purple. The foot stalk is about 3 inches in circumference, and is covered with soft fleshy spines. The root of the *daiga* is ripe when the leaf dies down, in which condition it remains until the flower appears above the ground the next season. It is considered nutritious although very acrid, and it cannot be used until cooked. It is said to aid fermentation, and is mixed with some other roots and fruit in making Fijian bread *madrai*.

Another esculent root *yaka*, or *wa yaka* (pachyrrhizus angulatus), is also found wild in all parts of Fiji, but not so common on wooded as on open grass land. It is a herbaceous plant, bearing trifoliate leaves and white flowers tinged with blue. The roots, or tubers, are the edible part of the plant. They grow to a size of about 3 feet in length and 3 or 4 inches in diameter. They have a flat, sweetish taste, and are very stringy even when well cooked.

The *tavoli* (dioscorea nummularia), yields an esculent root which, when in season, is much relished by the Fijians. It grows wild in the forests, and abounds in some parts of Taviuni, Ovalau, &c. The root is about the size of a large carrot. It is generally eaten roasted, and, although a little fibrous, is dry and mealy, and its taste is not unpleasant. The *kaili* (dioscorea bulbifera) is another yam, which is found growing wild in the woods. It is not so much relished as the *tavoli*, and requires to be soaked in water to remove its acridness before being boiled or roasted.

The *masawe* (dracæna sp.) is largely cultivated by the Fijians, and may be seen growing round the houses in every native town. The root is large and soft, and full of a sweetish juice resembling stick-liquorice in taste. It is either chewed by itself, or used by the natives to sweeten their *vakalolo* —puddings.

The cassava, or tapioca plant (jatropha manihot), has been recently introduced. Single plants and occasionally small plantations of it were noticed in the vicinity of several native towns, and it was seen growing like a wild plant in many places in the forests. Its presence there is most probably owing to cuttings of it having dropped out of a bundle while being carried through the forest. It grows readily from cuttings of the stem. These cuttings are generally about a foot in length, and are laid flat in the ground at the bottom of a hole about 6 inches deep, and then covered with soil to the depth of an inch. As an article of food, the root is not much relished in Fiji either by the natives or Europeans. In Seychelles it is otherwise. There the roots are frequently eaten raw, and by law, the planters, supply them for food to the labourers on the cocoa-nut plantations. The young roots when boiled are served as a vegetable and are very palatable, but become hard and fibrous as they grow old. The roots are also grated into a kind of meal, from which cakes are made (much the same as oatcakes in Scotland). These cakes are to be found on almost every table in Seychelles.

The *ivi* nut, in its season, is also largely used as food throughout Fiji. When grated, it forms one of the ingredients of Fijian *madrai*. The kernel is roasted or boiled in the pod, and eaten either cold or hot. Its taste is not unlike that of the chestnut, but to some stomachs *ivi* nuts are very indigestible, even when well masticated. The nuts of the *tavola* (terminalia catappa) the "Country almond" of India, is also extensively eaten during its season. It requires no cooking, and has a pleasant almond-like taste.

In order to preserve the produce of the earth from one season to another, or to guard against want during a season of drought, or a scarcity of food from any cause, the Fijian digs a circular hole in the ground 4 or 5 feet deep, and lines it with leaves. The hole is then filled with bread fruit, bananas, *dalo, daiga, kawai*, a kind of yam, &c., all thrown together or placed in layers. The whole is then thickly covered with leaves and earth. The mass soon begins to ferment, and emits a strong, sour, and very nauseous smell, which may be felt at a distance of more than half a mile on the leeward side of the pits. In a short time the different materials form a tough doughy looking lump. A portion of this is taken out as required, and baked on hot stones or steamed in an earthen pot. Such is the materials from which the Fijian bread, *madrai*, is made,—certainly not inviting to the palate of a European. In this manner some kinds of food are kept from one year to another.

So few potatoes are cultivated in Fiji, either by the natives or by settlers, that they are scarcely worth mentioning. That they would grow well, especially in the dry parts of Fiji, and during the dry and cool season, cannot be doubted. The settlers have not time to devote to their cultivation, and the natives are still ignorant of it. At present small quantities of them are imported from Australia and New Zealand, chiefly by dealers, and sold to the settlers.

As regards European culinary vegetables, Fiji is badly supplied. In fact vegetables grown in the colony are scarce and of bad quality, and seldom found on the table of any of the settlers. One or two market gardens have been established by Chinamen in the vicinity of Levuka ; but even there, where there is more demand for vegetables than in any other place in Fiji, the supply is meagre, and neither fruit nor vegetable markets exist. From the gardens of these Chinamen, cabbages, lettuce, parsley, shallots, radishes, French beans, pumpkins, and sometimes carrots, turnips, and cucumbers may be obtained, but the quality is invariably bad. This is owing to bad cultivation,—not supplying sufficient water to the growing vegetables, and cultivating them at wrong seasons. This will be remedied in course of time, as the demand becomes greater, and the people get acquainted with the method of cultivating the different kinds, the best season for planting, and the kind of soil best adapted for each.

Between vegetables peculiar to the tropics in the warm wet season, and those belonging to the temperate regions in the cool dry season, Fiji ought to be well supplied with salads, pot vegetables, and fruits of home growth all the year round.

Parsnips, rhubarb, long carrots, brussels-sprouts, and brocoli will not likely succeed well in Fiji. But turnips, cauliflowers, celery, kohl-rabbi, salsifa, red cabbage, broad beans, would succeed well during from three to six months of the cool season of the year. Beet-root, green peas, spinach, cucumbers, pumpkins, endive, chicory, lettuce, turnips, and radishes, could be on the table during from six to nine months of the year ; while potatoes, asparagus, cabbage, short carrots, onions, radishes, green maize, tomatoes, brinjals, french beans, water cresses, ought to be plentiful during from nine months to all the year. The want of any one of these vegetables during the above periods, of good quality and cheap, must not be attributed to either the soil or climate of Fiji. In New Caledonia, and Mauritius, where soil and climate closely resemble those of Fiji, all the above-named vegetables grow during the periods mentioned respectively.

The Fijians have of late taken to the cultivation of some European kitchen vegetables, notably cabbages, which they hawk on the streets of Levuka. A small garden planted with them was seen at na Sau-coka, in the interior of Viti Levu. The garden belonged to the native minister of the " Koro."

Were the Fijians granted a supply of vegetable seeds to begin with, and taught how to cultivate each different kind, it may justly be inferred from the liking which they have for field and garden work, that they would soon be able to supply the community with good and cheap vegetables.

The Fijians are fond of green vegetables, and cultivate several plants to supply them, in addition to those that grow wild. They generally eat them with pork and fowls, or fish. In heathen times, several kinds were eaten with *bakolo*, or human flesh. Notably the *malawaci* (trophis anthropophagorum), *ludano* (omalanthus pedicellatus), *borodina* (solanum anthropophagorum), and *kurilagi* (a peculiar variety of *dalo* (colocasia antiquorum). The vegetables commonly used are the young leaves of the *dalo*. Like the root they are very acrid, and require to be well cooked, in order to remove their acridity and make them wholesome. The *boro ni yaloka ni gala* (solanum nigrum, var. oleraceum), is also used by the Fijians, but not much by the settlers. It generally grows on cultivated lands. The *bele* or *vau vau ni viti* (hibiscus manihot), is much used and extensively cultivated in several districts by the natives. It has a taste not unlike spinach.

At Rabi, the leaves of a plant belonging to Phytolaccaceae were cooked and served as a vegetable daily at dinner. They are superior to the *bele*, and equal to spinach. Indeed, some consider them superior to it. *Taukuku ni vuaka* (portulaca oleracea), a weed growing on all waste land, is also much used by the natives. The young undeveloped flowers of the ⁻*vaulo* (flagellaria indica), a gramineous climber common in all forests throughout the colony, are also in great request during their season. Another gramina, the *drauka*, a plant somewhat resembling the sugar cane, is cultivated largely in some parts of Fiji. As a vegetable it is much relished by the Fijians all over the group. The unexpanded panicle of young flowers is the part eaten. If taken when young and tender, properly cooked, and served with butter as sauce, it is reckoned, by some, not inferior to asparagus. I regret that my specimens of this plant were not in fit condition to be named. They were not sufficiently advanced, and from the demand for the flowering shoots, specimens in full flower could not be obtained. To obtain these in Fiji, a *tabu* or prohibition to touch, would require to be put on a few plants. Besides the above plants, the young leaves of several kinds of ferns are used as pot vegetables. These are litobrochia incisa, alsophila excelsa, or *bala-bala*, in times of scarcity, angiopteris erecta, and asplenium esculentum—*oto*. The last, from choice, is used most extensively at all seasons and on all occasions. In general, the Fijians boil the vegetables, drain off the water, and serve them in the same way as spinach. Sometimes they are baked along with pork, fowls, or fish, at other times they are used as ingredients in fowl or fish soup. These soups are often well prepared and nicely flavoured.

CHAPTER IV.

FRUIT.

Fruit is plentiful in Fiji. Bananas, as already mentioned, abound everywhere. Of these musa troglodytarum, *soaga*, is said to be found wild in the woods. I saw it only in plantations, or in land which had been cultivated. As an article of export, to the Australian colonies and New Zealand, bananas deserve to be extensively cultivated. Practically, the market for them is unlimited, and their cultivation, which is not attended with much expense, will be remunerative. The same may be said of pine apples, the *balawa ni papalagi* (or foreign pandanus) of the Fijians, which thrive remarkably well in Fiji. The soil and climate are in a high degree suitable for their cultivation, and they are produced in perfection. The demand for them in the Australian and New Zealand markets is great; the supply is scanty and unequal to the demand. A few are grown by the Fijians, but the cultivation of them is by no means general throughout the group,—the gardens of the settlers excepted. The cultivation of the pine apple and good bananas, with perhaps oranges, limes, and lemons by the natives, at easy distances from the ports where the colonial steamers call, should be encouraged by Government, which might take the fruit from the natives as part of their taxes. Two or three varieties of the saddock, *moli kana*, have extended to all parts of Fiji; so also has the lemon, *moli kuru*

kuru, and the lime, but the latter is not so plentiful as the two others. *Moli kuru kuru* was seen at some places planted as a "*fighting fence*" around some of the native towns. In many places the two former, in a wild state, are more common than any other kind of trees, and especially so between Suva and Kaluba, where the air was loaded with the smell of the decaying fruit.

The orange, *moli ni tahiti*, is not so common as any of the above, still, it is not rare. Rewa is noted for the excellent quality of the oranges, which used to be exported to Australia before they began to be extensively cultivated in New South Wales. As the orange ripens in Fiji at a different season from Australia, it might still be exported there, and especially to New Zealand, where the climate is too cold to grow them in the open air. Auckland, New Zealand, is about as near Suva and Levuka in Fiji as it is Sydney, and from the climate and cheapness of labour, oranges could be grown at a less expense in Fiji than in Australia. At Namosi, in the interior of Viti Levu, oranges are abundant, and their quality cannot be excelled by those of any country. Dr. Seeman left seeds of the orange there with Mr. Danford, known in Fiji as "Harry the Jew," and gave him instructions as to the treatment the trees required. The healthy look of the trees, and the large crops of fruit that they bear annually, give evidence that these lessons have not been neglected. The appearance of young trees in other districts, particularly in the neighbourhood of Suva, indicates plentiful crops of oranges in future years. But to insure this result some attention must be given to the trees, which, like some other things in Fiji, are generally left to nature to grow or die. If the shaddock and lemon are not indigenous to Fiji, it is difficult to say when, or by whom they were introduced. Large trees of the former, with trunks from a foot to 18 inches in diameter, may be seen in all parts of Fiji, even in the centres of supposed virgin forests. Although lemons and limes abound in Fiji, no lime juice is made, but the abundance of the raw material suggests the making of that commodity largely and cheaply.

The *dawa* (nephelium pinnatum), is a kind of litchi, but its fruit is inferior in quality to that of the latter. When ripe it is about the size of a pigeon's egg, not unlike it in shape, of a yellowish green colour, and covered with soft fleshy spines. The taste is flat, sweet, but not unpleasant. The aril, surrounding the seed and lying between it and the skin, is the part eaten. In some varieties this part is about three-eighths of an inch in thickness. It is soft and glutinous, and has a pale yellow semi-transparent colour. There are many varieties of the tree, and good hopes may be entertained that the quality of the fruit will be improved by careful cultivation. The fruit of some varieties of the *dawa* is nearly twice the size of the litchi; and a fruit bearing variety, combining the size of the *dawa* with the flavour of the litchi, would be an important addition to tropical fruit. Owing to the seeds germinating readily wherever dropped, forests of the *dawa* may be seen in many parts of Fiji,—most commonly on land that has once been cultivated. The reason for this is that the growth of these young trees has not been prevented by that of other trees previously established on the soil.

Eugenia malaccensis (Malay apple), the *kavika* of the Fijians, is another indigenous fruit. There are several varieties of it, but those most esteemed are the white and the red fruit bearing kinds. It is a fine symmetrical growing tree, frequently attaining a height of 40 feet, and its dark green foliage contrasts beautifully with the bright red colour of its large flowers and fruit. The fruit is borne on the branches in clusters of three or more together. It has a flavourless watery taste to a European, but is much relished by the natives, who besides eating the fruit, use it as an ingredient in their *madrai*, and the flowers as an ornament for their hair.

The *wi* (sponias dulcis) also abounds in many parts, and is most common about the native towns. It is a sparsely and horizontally branched, deciduous tree, frequently growing to a height of 70 feet. The fruit, which is much relished both by the natives and Europeans, is of a yellow colour when ripe, very juicy, and has a pleasant sub-acid flavour. It is about 8 or 9 inches in circumference, and sometimes weighs nearly a pound when well grown, but it is full of wiry fibres that make it disagreeable to persons who have tender gums. The settlers make it into tarts, and also eat it in its natural state. The tree is more common, grows better, and its fruits have a finer flavour on the coast than on the elevated land and wet climate of the mountains.

The *tarawau* (dracontomelon sylvestre) is another Fijian fruit bearing tree. It grows to a height of 40 feet, and has a wide spreading head of branches with bright green pinnated leaves. The fruit is succulent and juicy, but the taste is flat and wanting in flavour. It is relished only by the natives.

Planting the *larawau* is, according to Fijian mythology, the occupation of the departed in a future state.

In addition to these, there are several other kinds of fruit eaten by the Fijians, but not much relished by Europeans. They are the *lose lose* (ficus vitiensis), the *balawa* (pandanus caricosus), the wild bramble *wagodra-godra* (rubus tiliaceous), the *bakoi* (eugenia richii), the *sea* (eugenia sp.), the *nawa-nawa* (cordia subcordata), the *vutu kana* (barringtonia excelsa); in fact, to a hungry Fijian scarcely any kind of fruit, animal, bird, fish, or snake comes amiss.

Water melons are abundant in Fiji, and are much esteemed both by Fijians and settlers. The bottle gourd is also common, but not the sweet melon, which, from some cause, is entirely neglected. The carica papaya (the papaw, mamme apple, or mummy apple,—as the settlers wrongfully persist in calling it), is not uncommon throughout the group. The Fijians eat it raw, but by Europeans it is cooked and eaten as a vegetable or a preserve. "The leaves possess the " property of making tough meat tender; and the " seeds are an efficacious vermifuge for children."—(Seemann.)

Of other exotic fruits that have been introduced into Fiji, and thrive there, may be mentioned the following, viz., the guavas, psidiums pomiferum, pyriferum, chinensis, and cattleyanum. The first two are now quite naturalised in Viti. They are most common in the vicinity of European settlements, whence they will ultimately spread throughout the colony. On the rocky soils near Levuka they have grown into a dense scrub in some places. The last-named two are of recent introduction, and it is not long since they began to bear fruit, but they will soon be equally at home. The little black-fruited psidium cattleyanum resembles the strawberry in taste, and is the best flavoured of the four. The yellow fruit of psidium chinensis is the least palatable, but it makes an excellent preserve. Guava jelly is made from the first two, but there is not so much of it made in Fiji as in Mauritius and some other countries. The *loquat*, or *bibassier*, has also been introduced, but at present it is only grown on the coast, where the climate is too hot for it bearing well. In the tropics it bears best in elevated situations, where the climate almost resembles that of temperate regions.

The white and the black fruited varieties of the mulberry (morus indica) are common throughout Fiji. Some towns in the interior of Viti Levu are surrounded with a fence of them. A bank, or wall of earth and stones, has been thrown up from 4 to 6 feet in height as a sort of fortification, with a broad and deep ditch on the outside (frequently on both sides of the bank). On the top of this wall or bank, a thick fence of mulberries, lemons, and, in some places, of barringtonia racemosa and jatropha curcas, has been planted. The two last chiefly at the Koros on the coast or lowlands. The hollow trunks of tree ferns have been laid horizontally through the wall, and used as loopholes for the defenders to fire from. The healthy, thriving appearance of the mulberry in Fiji suggests that silk, in the form of cocoons, might be added to the products of the colony. No doubt need be entertained about the suitability of the climate of Fiji for rearing the silkworm, and the cocoons could be produced very cheaply. "The growth of silk" among the natives might be encouraged by Government.

The anona squamosa, or custard apple, thrives well in Fiji, and is now common in the gardens of the settlers. This augurs well for the success of other fruit-bearing species of anona.

The grandilla (passiflora quadrangularis) is quite common, and bears annually loads of delicious fruit.

The pomegranate is not less common, but it does not bear so freely. The rind is a reputed cure for dysentery in some countries.

The mango has also been introduced, but so recently that only a few of the older trees have begun to bear fruit. No certain idea can be formed as to how the flavour of the fruit will be affected by the climate of Fiji. The young trees present a hopeful appearance. The mango is more prolific and the fruit is of better quality when grown in a dry climate than in a wet one, notwithstanding that the trees are generally larger and healthier in the latter than in the former. Moist, showery weather is more favourable to the growth of wood than of fruit. Although in a wet district the trees may yearly show an abundance of flower, yet, should a tropical shower pass over them while in blossom, scarcely a flower will set in a whole orchard of mango trees. Hence it may safely be affirmed that the best crops and the finest flavoured fruit will be produced in the dry localities of Fiji. The prevalence of either wet weather, or a cloudy sky, prevents fruit from being well flavoured in any country.

The alligator pear or avocada (persea gratissima) has lately been introduced, and its success may be deemed certain. The same may be predicted of the jack, tamarind, litchi, longan, &c. Regarding the mangosteen and durian, one cannot speak with the same certainty, as they are natives of a much hotter climate than Fiji. However, a trial of both should be made, and were the trees planted in sheltered valleys at a low elevation, where the climate is moist and warm, the soil rich, constantly moist, but well drained, the results would be hopeful.

Moderate success has hitherto attended the cultivation of the fig tree (ficus carica) in Fiji. It is in every settler's garden, and frequently bears a fair crop, but the fruit is not so well flavoured as in a temperate climate.

Considering that some varieties of the peach do well in Mauritius and New Caledonia, there can be no doubt that they will grow equally well in Fiji, especially if the varieties are obtained from either of these colonies, or the northern parts of Queensland.

Strawberries also grow fairly well in the above islands, and it may be expected that they will prosper in Fiji, particularly in the cool parts of Viti Levu, Vanua Levu, and the mountains of Taviuni and Ovalau. The climate of Fiji is in no part cold enough for the common raspberry, but another species (rubus rosæfolius), common in the forests of Mauritius and Seychelles, would thrive well in Fiji, and prove a valuable addition to the fruit-bearing plants of these islands.

The climate of Fiji is not favourable to the cultivation of the grape-vine. The plants grow fairly, become evergreen, and occasionally bear a few bunches of small fruit. The climate is not too hot for the vine, and the want of success is due to the plant not getting a season for rest. This might be remedied by planting it in those parts where rain seldom falls; the drought would have the same effect on the plant as the cold weather of a temperate climate. However, the cultivation would require to be carried on entirely by irrigation, and the roots kept under the complete control of the cultivator. In this way two or three crops could be got in the year from the same plants, as at Jaffna, Ceylon.

Such fruits as the apricot, plum, cherry, apple, pear, gooseberry, currant, &c., will not grow in Fiji, as the climate is far too hot for them.

CHAPTER V.

ORNAMENTAL PLANTS—STARCH—SPICES—CLOTHING
—MATS—FANS—CORDAGE.

IN the vicinity of their dwellings, and in their burying grounds, the Fijians cultivate several kinds of flowering and foliage plants. Of the latter were seen several kinds of fine crotons, 20 varieties of beautiful dracœnas, an immense number of coleus of nearly all colours, the admirable amaranthus salicifolius or tricolor, the pretty alternanthera paronychoides (in several varieties), called by the settlers "Fiji grass," three or four kinds of acalyphas, two or three varieties of panax, near p. fruticosus, *dani-dani*, one or two species of euphorbia, a dwarf variety of hibiscus tricuspis, *vaudra*, two kinds of ornamental grasses, one having reddish, and the other variegated (green and white) leaves. Among flowering plants were noticed some gay varieties of hibiscus, near h. rosa-sinensis. Several double flowering sorts of hibiscus are cultivated; one, with bright scarlet flowers about 6 inches in diameter, was very conspicuous. Balsamins, gomphrenas, and *marvel of Peru*, are common favourites of the natives. The fragrant white trumpet flowering brugmansia suaveolens, a few varieties of roses, bauhinia richardsoniana, &c., are also much coveted by the Fijians. The natives like to have sweet smelling flowers growing near their houses, such as the *bua* (fagræa berteriana), one or more species of drymispermum and leucosmia, locally termed *sinu dina, sinu danu*, and *mataiavai*, and the peculiar scented *uci salu-salu* (evodia hortensis), whose leaves when crushed emit a pungent odour resembling peppermint. With the flowers of the *bua, sinu dina, mataiavia, makosoi* (cananga odorata), *bua siu* (lindenia vitiensis), some species of dolicholobium, *buabua* (guettarda speciosa), jasminums, *wa vatu*, the *vasa* (cerbera lactura), and hoyas, *wa-bibi*, the Fijians make wreaths, necklaces, sashes, and belts or bands. The sashes go over the shoulder, cross the chest, and pass under the arm, and the belts are worn round the body. The flowers are strung upon a cord, made from the bark of the *vau* (hibiscus tiliaceus), or a bit of a tough climbing plant, and the corolla tube of one flower is inserted into the corolla tube of the flower next below it. Sometimes chaplets wreaths, &c., are made of the *wa kalo* (lygodium scandens), and the fronds of other delicate ferns, artistically woven or plaited together, along with the red leaves of dracænas, and those of odoriferous plants as the *cevuga* (amomum augustifolium), or a graceful mixture of foliage and flowers, is not unfrequently used.

The native, in ornamenting himself with gay flowers, seems to study his own taste, and takes delight in being adorned differently from his neighbour.

They also profusely anoint their heads and bodies with oil to which a fragrance has been imparted by the gratings of sandalwood, *macoua*, the bark of cinnamomum pedatinervum, and some other species of the same genus;—The flowers of the *maketa* (parinarium laurinium), *makosoi, laga kali* (my specimens of this plant were not sufficient to identify it), *lebu* (eugenia neurocalyx), the leaves of a species of litsea, the *uci*, &c. In a word, everything is used that will impart a perfume to the oil, but the odours of sandalwood and *macou* are the most valued.

In the gardens of European settlers many tropical flowering shrubs and trees, climbers, and ornamental leaved plants are cultivated; by some with care, taste, and skill. These plants are chiefly cletorias, hibiscuses, cranthemums, russellias, ipomæas, dracænas, acalyphas, coleus, alternantheras, gardenias, roses, &c. Every vessel from Australia, New Hebrides, the Solomon, or Marquesas islands, brings additions to the collections of amateurs, such as rare specimens of hibiscus, dracænas, crotons, palms, &c.

The Fijians do not make use of the *roto* (cycas circinalis) or the *soga* (sagus vitiensis) as sago yielding plants. This is an article they seem to be unacquainted with, and it has not been made in Fiji from these plants by any one, Seemann excepted. From the roots of the *yabia* (one or more species of tacca), a fecula is made. It is like arrowroot, but darker in colour, and more nutritive. It is used in the same way as arrowroot. Those who have tried both prefer the *yabia* when they can obtain it pure and clean. The *yabia* grows abundantly in all parts of Fiji, but it is more common in the interior of Vanua Levu than any other locality visited. The plant is not cultivated, but grows freely among grass in almost any soil. It is herbaceous, and the leaves come up annually after the wet season has set in. At the beginning of the cool dry season the leaves turn yellow and decay, but the stalks remain in the ground, free from the tubers. At this stage the tubers are in proper condition for digging, and the decayed leaf stalks point out their exact position to the diggers.

Turmeric (curcuma longa), the *cago* of the Fijians, abounds in all the islands. Like the *yabia*, it is herbaceous, and the rhizomes are fit for digging when the leaves have died. When prepared, it is called *rerega*, and is used for cosmetic purposes, dyeing garments, &c.

Cassava (jatropha manihot), is cultivated by some of the settlers for making tapioca, but there has not yet been much of it made. The "plant" for grating the roots and granulating the tapioca, is but recently set up.

Arrowroot of very fine quality is made in one of the islands, from maranta arundinacea, and a species of canna which has recently been introduced. Both plants thrive remarkably well, but the largest yield of fecula is obtained from the canna. A considerable area of land on the island of Koro has been planted with these, and machinery for the manufacture of arrowroot has been erected there by an enterprising gentleman. Steam power is used for grating the rhizomes or roots, and the fecula is dried by heated air. The estimated yield per acre is said to be about 1,300 lbs. The arrowroot is sent to England, Australia, and New Zealand. It is cleanly made, pure, and its quality is not inferior to that made in any country. Some persons think that the quality of white arrowroot is inferior to that which has a slight brown or yellowish tinge. This is a mistake. Arrowroot when pure is white; when coloured in the slightest degree, it has not been perfectly cleaned. As the quality of white sugar is superior from its cleanness to brown, so the quality of white arrowroot is superior to that which is coloured. The **value of the arrowroot exported during 1878 was 776*l*., in 1877, 500*l*., and 281*l*. in 1876.**

Three or four species of nutmegs (myristica), called *male*, grow wild in the forests of Fiji. Some of the settlers use the fruit of one of these (myristica castanæfolia) as a substitute for the nutmeg (myristica moschata). It closely resembles the latter in size and shape, and in the colour of its aril, or mace, but its aromatic properties are not so well developed. A quantity of them was sent to Sydney, but the Chinese in that city soon detected the want of flavour, and there was no further demand for FIJI NUTMEGS. It may be presumed, from species of the genus being found in the colony, that the soil and climate will prove favourable to the growth of the true nutmeg. They are undoubtedly favourable. Although proof of success is yet wanting, no one need hesitate in planting the tree in Fiji, especially in the sheltered valleys and hillsides in the interior of the large islands.

None of the native cinnamons possess the grateful aromatic fragrance of the cinnamon of Ceylon (cinnamomum zeylanicum), and, consequently, they are of no value as articles of commerce. Their presence, however, among the indigenous plants of the colony, bespeaks a successful issue in the cultivation of the true species whenever it may be tried. The cinnamon is a hardy plant, which will thrive in the poorest soils. There is no doubt of its growing well in Fiji. By some of the settlers *macou* is used as a substitute for cinnamon in seasoning custards, puddings, &c., but, at the best, it is a poor exchange.

The clove and allspice will thrive in all parts of the colony, the soil and climate being favourable to them. There are no seed-bearing spice trees in Fiji, and however well suited these islands may naturally be for their growth, the cultivation of them will not be extensive until the trees bear seeds in the colony. No doubt this is the case with every new species of plant that is introduced. But while the plants are growing time is not lost. It gives intending planters leisure to think. Many erroneous ideas are exploded; a great deal of information and experience is gained as to the kind of cultivation the newly-introduced plants require; kind of soil and situation that suits them best; most efficient method of manufacturing or preparing the produce for market; and in this way the enterprise is taken beyond the region of hazardous speculation or experiment.

The cultivation of ginger (zingiber officinalis) has lately been tried in Fiji. The plant will thrive; but as yet planters are only feeling their way as to its proper treatment, and the preparation of the produce for market. There is a native ginger (zingiber zerumbet), but it is of no market value, as its rhizomes have an unpleasant flavour.

Several species of piper (pepper plant), are found wild in Fiji. None of them yield a product of any value. The colony is eminently adapted for the cultivation of the pepper plant (piper nigrum). A few plants of it have been introduced, and although the stock can be easily increased by cuttings, ignorance as to the cultivation of the plant, and the manner of preparing the berries for market, will be a draw-back for some time to come.

Piper methysticum *agona* is grown to some extent by the natives (chiefly round their houses), for making their favourite beverage *agona* or *kava*. When well grown it is a noble, picturesque-looking plant, worthy of cultivation in our English hot-houses. The root is the part used, and the plant is easily propagated by cuttings. Of late, attention has been called to the medicinal properties which the plant is said to possess, and a quantity of it has been exported to the Australian colonies. The Fijians give presents of its roots to their chiefs, friends, and anyone that they wish to propitiate. Between them and some of the traders it has a recognised market value, at which it is bought and sold, varying according to the supply and demand. The best quality is grown in the mountains where it grows most luxuriantly, and in great abundance. The Fijians say that the bird's-eye pepper or chillie plant, *boro ni papalagi*, or foreign boro (capsicum frutescens), is not a native of their country. If so, it has long been naturalised, and is now one of the most common plants in these islands. It bears an abundant crop of fruit, very pungent in quality, from which excellent cayenne pepper could be made. It is largely used by the settlers to give pungency to curries, sauces, &c. Some kinds of domestic fowls, such as turkeys, guinea fowls, &c., are very fond of its fruit, and it is said to give their flesh a pleasant spicy flavour.

Among the natives, the use of cotton cloth for *sulus* is annually increasing. Still, their wants in this respect are to a great extent supplied by the bark of the *masi* (broussonetia papyrifera), which is extensively cultivated for the purpose. The mucilage, &c., is beaten out of the bark with wooden mallets; and the glutinous nature of the fibre serves in a great measure to join pieces of the bark together till the required size is attained. To strengthen the joinings, a paste made of *yabia*, or some other adhesive substance, is used. The CLOTH is sometimes made to a degree of fineness resembling the finest gauze; and by being frequently washed and bleached it becomes perfectly white.

Besides *sulus* and turbans for the head, mosquito curtains and hanging screens are made of it. These screens are used as partitions to divide a house into compartments. They are printed in various patterns which, although displaying a good deal of taste in design, are poorly executed. The colours are placed on raised forms made of strips of wood, generally of bamboo. The juice of the *lauci* (aleurites triloba), *dogo* (rhizophora mucronata), and *kura* (morinda citrifolia) is largely used for this purpose.

A coarse cloth is made, in the same manner as *masi*, from the bark of several species of *baka* (ficus), which grow in great numbers in Fiji. It is worn by the poorest class of the people, and only, when better cannot be got, by the well-to-do.

The *liku* is a dress much worn by the Fijians on festive occasions. It is made of the fibres of different kinds of plants. One end of the fibres is fastened to a band which is tied round the waist, the other end hangs down over the thighs like a fringe. The length of the *liku* varies according to the rank of the wearer, some of them reaching to the knees, others only to the thighs. The most fashionable *likus* are made from the *wa loa*, a species of fungus (rhizomorpha), that grows on decaying trees in swamps, and prepared leaflets of the cocoa-nut tree stained black or yellow by mud and turmeric respectively. The fibres extracted from several species of *vau* (paritium tiliaceum, tricuspis, and purpurescens), are are also used for *likus*, particularly in the interior of Viti Levu. The tough woody branches and stems of some climbers, beaten into fibres, are used for the same purpose. The climber from which the most of these are made is alyxia bracteolosa.

When working, or travelling along a muddy path, or through long wet grass or reeds, the Fijian doffs his clean cotton *sulu* or fashionable *liku*, and dons one of the latter, made of very simple material, viz., one or two banana leaves, which he fastens round the loins by their mid-ribs and allows the blades of the leaves to fall in shreds over his thighs, A man's *liku* hanging on a stick or laid upon a stone by the side of a stream, is a *tabu* against women bathing there; and if a woman's *liku*, it is a *tabu* against any man going near the place. Anyone who knowingly breaks the *tabu* smarts for it.

The leaves of some species of pandanus (p. odoratissimus and caricosus) *balawa* and *voivoi*, are used for making mats to cover the floors and sleep upon. The finest mats are made from two species of sedge, called *kuula* and *ya* (eleocharis articulata and variegata) which grow profusely in swamps. The leaves are well bleached in the sun before being used. This renders them tough and pliable, and gives the mats a yellowish colour when made up.

Fans are made from the fibres of various plants. Those most valued are made from a single leaf of a fan palm (pritchardia pacifica), bordered with a strip of bamboo, to prevent the leaf splitting. The borders of mats and fans are tastefully ornamented with coloured wool, feathers, &c.

Besides for making mats, the above materials are also employed in making baskets of divers models. Different designs are wrought out on the baskets, by using plain or stained material. The strongest baskets are made from the split stems of flagellaria indica, and the mid-ribs of cocoa-nut leaves; a strip from the outer part of these being used. A kind of ratan (calamus), well adapted for basket making and all kinds of wicker work, is common in the mountain forests of Taviuni. As no articles made from it were seen, it is not supposed to be used for basket making, or in fact for any purpose. It is a strong growing plant, climbing to the tops of the highest trees. Its leaves, divided into leaflets like those of a palm, end in a spiny tendril, and their mid-ribs and veins are covered with recurved spines. The stems can be divested of the leaf-sheaths, which adhere to them, by steeping in water and exposure to strong sunshine.

Ropes, &c., are made from the fibres of *vau* (paritium sp.), *kalakakaisau* (hibiscus diversifolius), *sinu mataiavi* (wikstrœmia indica), *yaka* (pachyrrhizus angulatus), cocoa-nut fibre or *sinnet*, &c. These fibres are extracted by maceration, beating, or scraping with a shell. Fishing nets are principally made from the fibres of *sinu mataiavi*, and the *yaka*, plaited hard like whip-cord. They are as strong and lasting as those made of hemp. Plaiting *sinnet* is a favourite occupation of the Fijians, in which all can take part.

CHAPTER VI.

Timber—Native Houses—Minor Forest Produce.

The forest of Fiji contain many valuable timber trees, of which a few may be mentioned.

The *vesi* (afzelia bijuga), yields perhaps the most valuable timber in Fiji. It generally grows on the shore or sandy beaches, and in rocky clefts, and by the sides of streams in the interior (on the north-west sides) of Viti Levu and Vanua Levu. It is not a lofty growing tree, but its branches are wide spreading. The length of the trunk is seldom more than 20 feet, and its diameter is frequently from 3 to 4 feet. Its wood is of a dark brown colour, closely and evenly grained, heavy, and hard, but easily worked; very durable in any situation, and capable of taking a fine polish. It is a most valuable timber where strength and durability are required, and in no respect inferior to teak. Knotty and gnarled portions of it are excellent substitutes for lignum vitæ in making block-sheaves. Large sound trees of it are now scarce in the colony, owing to its having been extensively used in the construction of large canoes, for which the Fijians were famed. It is used in house building as pillars to support the wall plates and centre ridges; also in making *agona* bowls, native drums, or *lalis*, &c. The tree is common on the shores of nearly all tropical countries, from the east coast of Africa, through the Mascarene, Indian, Malayan, and Polynesian islands to the western shores of America. But it is not common, if at all found indigenous, at a distance of more than 20 degrees of latitude north or south of the equator. It is not found wild in Mauritius, and I did not see it in the Sandwich islands. It is of slow growth, requiring about 140 years, or more, to reach maturity.

The *tavola* (terminalia catappa) is found like the *vesi* over a wide range of tropical countries, either growing spontaneously, or planted for the pleasant shade of its foliage. In Fiji, it grows naturally on the shores, and on the banks of streams, in the interior of the larger islands. It is also planted about the native towns for its shade. It is a lofty horizontally branched, deciduous tree, which frequently grows to a height of 70 or 80 feet, with a trunk from 4 to 8 feet in diameter, and sometimes a height of 40 feet from the ground to the lowest branches. The timber is of a light brown colour, with darker veins, easily worked, not very hard nor heavy, and about the same density as teak, which, in outward appearance, it somewhat resembles. The timber is durable when kept dry, but decays fast in places where it is alternately wet and dry. It is very useful for house building purposes, such as flooring, partitions, inside doors, &c. The *tavola* is said to make the best sounding *lalis*. These are dug or hollowed out, like a trough from a portion of the trunk, leaving thin sides, and about 4 inches of the wood at each end. The sides are beaten by two sticks about 18 inches long. In different ways, measures or time is beaten on them, the meaning of which is known by the natives and the "old hands" or settlers. The sound of the *lali* is not unpleasant when beaten by a practised hand, and a good sounding one may be heard at a distance of

4 or 5 miles on a quiet morning. The trunk of the *tavola* and some other trees, notably the *dakua* and *damanu*, are used, faute de mieux, for making "dug out" canoes or *takias*, for navigating rivers and smooth water round the shores, and inside the reefs.

The *dilo* (calophyllum inophyllum) is another large tree. Like the preceding two, it is common on the shores of most tropical countries, and not unfrequently found in the interior, where, if in dense forests, it sometimes attains a height of 60 or 70 feet, —being forced up by encroachment and protection of surrounding trees. On the shores it seldom grows to a lofty tree, but has large wide-spreading branches, and a trunk frequently 7 feet in diameter. Like the other calophyllums mentioned below, it yields a valuable and useful timber. The crooked branches of the trees grown on the sea-shore, are useful as "knees" in the construction of boats and large sized vessels. The seeds yield the *dilo* or "Tacamahaca" oil of commerce, which used as a liniment, is of high repute in the east as a cure for rheumatics, &c. When the bark of the tree is wounded, the sap, which flows gently from the wound, is of a clear, transparent, greenish colour. By exposure, it gradually thickens into a gum-resin which is the "Tacamahaca" resin of druggists. The oil as well as the resin is used by the Fijians and "old settlers" as a cure for rheumatics, pains in the joints, and as a balsam for wounds. There is not much of its oil now made in Fiji, and it is not easily obtained pure. The native method of extracting it is rude. As the tree is not uncommon in several localities, its seed might be collected by the people as a Government tax, and the market value of the oil ascertained. It is believed to be still of considerable value, and about 18 years ago it was worth 90*l*. per tun.

The *damanu* (calophyllum burmanni), *damanu dilo dilo*, and another unnamed species, yield the "*damanu* timber" of Fiji. It is of a light brown colour, beautifully but somewhat coarsely veined, tough, strong, and bears a heavy strain. It is deservedly esteemed in house and ship-building, being durable in mostly all situations. It warps, however, if not well seasoned, and the grain is rather curly, but it is not difficult to work. Planks and logs of it have been shipped to the Australian colonies, where they have met a ready sale at remunerative prices. Some of the timber has also been sent to the London market, but the result has not been made known. The timber of these species with that of the *dilo*, is celebrated for making excellent spars for small vessels, where strength and toughness are important requisites. The trees are common throughout Fiji, in dense forests on the windward sides of the islands, and near streams on the leeward or dry sides. They grow to a large size, and with what, in the tropics, is considered ordinary rapidity. The annual growth in height of a well-established young tree in fairly good soil will be from 3 to 4 feet. As the tree gets old the growth is slower, and it will probably take about 80 years to reach maturity, when it will have attained a height of about 60 or 70 feet, with a trunk,—a clean, smooth cylinder, of some 35 or 40 feet in length, having a diameter of $2\frac{1}{2}$ feet at the base and 2 feet at the top. The timber is useful for many purposes long before the tree has reached maturity, on which account the tree is especially valuable.

The *dakua* (dammara vitiensis) may be termed the *kawrie pine* of Fiji. It attains a great size, growing sometimes to a height of 100 feet, with a trunk of 27 feet in circumference and 60 feet to the first branches. This may be reckoned the maximum size of the tree. A living specimen that has almost reached these dimensions has been noticed. Dakua timber is light, close-grained, white, and easily worked. It makes good spars, and is useful for a variety of purposes. Its quality is equal to good pine. Large numbers of the best trees have been felled in the more accessible forests, and large-sized trees are now confined to the interior of Viti Levu and Vanua Levu. In some parts of these islands it is common, though not abundant, and it is now rare, if not extinct, in the other islands of the group. The tree exudes an inflammable gum-resin called *makadre* by the Fijians, who use it for burning as a lamp or torch. The gum is also used in glazing their pottery. The smoke of the burned resin also yields a black dye, which the Fijians used instead of the *lauci* in dyeing their cloth, tatooing, &c. The kawrie gum is a valuable article of commerce, but large pieces of it can now rarely be got in Fiji. The inflammable nature of the gum has undoubtedly been an important factor in the destruction of the trees by fire, which has exterminated them in many localities in the dry districts.

The *dakua salu salu*, or *kau solo*, as it is termed in some localities, (podocarpus vitiensis), is a fine handsome tree, in its leaves and appearance, closely resembling the Australian araucaria bidwellii. It does not attain the dimensions of the *dakva*, and seldom exceeds 60 feet in height,—its greatest circumference being about 15 feet. It is not so common as the dakua, but is found in Viti Levu and Vanua Levu forests from the level of the sea to the tops of the mountains. Its timber is of a light brown colour, easily worked, tough, strong, durable, and capable of taking a fine polish. It can be used for almost any purpose connected with house and boat building, and cabinet-making.

The *lewininini* (dacrydium alatum), is a smaller sized tree than the last named. It grows to a height of about 50 feet, and its trunk rarely exceeds 12 feet in circumference. It is found also in Viti Levu and Vanua Levu, and ranges from the sea-shore to the tops of the mountains. It yields a very useful timber, fit for almost every purpose connected with house building or cabinet-making. It is either white, or slightly cream coloured, close-grained, hard and durable, and is much valued by the natives.

Podocarpus cupressina (*kau tabua*, so called from the wood resembling whales' teeth in colour), yields timber of a beautiful yellow colour, closely and evenly grained, easily worked, durable, and moderately hard, useful alike for house-work and cabinet-making. It polishes well and easily. The tree grows in the forests of Vanua Levu and Viti Levu, and is found from the sea coast to the interior, but is more common on mountain ridges than in the valleys.

There are two other species of podocarpus, to which the Fijians give the name of *kausi*, viz., podocarpus bractelosa and affinis. They are trees of medium size, growing to a height of about 40 feet, with trunks about 6 feet in circumference. They are not uncommon, and generally grow on the tops of hills. They are seldom found in the valleys. Both species yield a fine timber, closely-grained, hard, durable, and useful for a variety of purposes.

The vai vai (scrianthes myraidenia) is a large tree. It grows to a height of about 70 feet, and has a trunk from 35 to 40 feet in length, with a circumference of 15 feet. It is common in the forests of the larger islands, and yields a strong, hard, white coloured timber, much valued in house and ship building on account of its toughness, strength, and durability. *takias* are sometimes made or dug out of its large trunks.

The *sagali* (lumnitzera coccinea), is a medium sized tree which grows to a height of about 40 feet, and its trunk girths about 7 feet, at 6 feet from the ground. It is much valued for piles, as its timber is durable in water and not subject to the attacks of insects. However, the trunk seldom exceeds 20 feet in height, which prevents its being used on all occasions. It grows in salt marshes, *tiri* or mangrove swamps near the mouths of rivers and on some parts of the coast.

The *kau kuru* (casuarina nodiflora) is most common in the district of Bua in Vanua Levu. There it grows to a fair sized tree of 50 feet or so in height, surmounted by a flat head of branches, and dark, sombre pine-looking leaves. Its trunk attains a girth of about 8 feet and a length of 35 or 40 feet. The wood is white coloured, closely-grained, hard and heavy. It is useful for many purposes, but does not bear exposure to the weather.

The *mulo mulo* (hibiscus, or thespesia populnea), a common tree on the shores of nearly all tropical countries. It seldom grows above 40 feet in height. The trunk is from 15 to 20 feet in length, with a girth of about 7 feet. The outside timber is white and soft like that of the willow, but the inside, or heart timber, is hard and durable. It is of a light-brown colour, curly-grained and not very easily worked, but a good timber for boat building and various other purposes. The large crooked branches are used as knees. In some tropical countries, such as Ceylon, this tree is extensively planted for its shade, along the sides of roads and streets. Large branches lopped off and put into the ground strike root rapidly, and in a few years are medium sized trees.

The *vuga* (nelitris vitiensis) and *vuga vuga* (metrosideros polymorpha) are trees of less than medium size. They grow to a height of 30 or 35 feet, with trunks about 20 feet in length, and seldom exceeding $2\frac{1}{2}$ or 3 feet in circumference. They are most common in the dry parts of Viti Levu and Vanua Levu, and grow in the poorest soil. Both are pretty flowering trees. The timber is very hard and durable and is used by the settlers as piles to support the framework of their wooden houses. For this purpose they are let into the ground to a depth of 1 or 2 feet, and the earth rammed hard about them to keep them firm. They are then all cut to the same level, at the required height above ground, for the floor of the house. In such a situation good timber of either kind is said to stand the vicissitudes of climate and weather for about 20 years.

The *koka* (Bischoffiia javanica) grows to a height of 45 or 50 feet, and has a large wide-spreading head, resembling the plane-tree in outline. The trunk is seldom more than 20 feet in length, and does not exceed 12 feet in girth. The tree is not often seen in thick forests, but is abundantly represented on land that has been cultivated, particularly on alluvial soil near the rivers. Orchids, ferns, and lycopods would seem to have a special favour for the *koka* as it may be seen covered with these epiphytes, while upon the surrounding trees there are none. The timber is not unlike teak in colour and grain,—hard, durable, and stands exposure to the weather. For general purposes it is quite as useful as teak, and the Fijians esteem it next to *vesi* in making posts or pillars for their houses. It would make excellent household furniture, and would polish well. When green, it is dense and heavy, and will not float in fresh water.

The *cibicibi*, cynometra sp., is a lofty tree, growing to a height of 75 feet, with a trunk 40 or 50 feet in length, and about 12 feet in girth. The heart wood is closely and finely grained, hard, durable, and useful for all domestic purposes. That on the outside is soft and white, and decomposes rapidly. The *cibicibi* is common in forests up to an elevation of 1,000 feet, but is most abundant in the wet districts, where it attains the greatest dimensions.

The *noko noko* (casuarina equisetifolia) is common in Fiji, but abounds most in dry localities. No particularly large trees of it were noticed, but in Seychelles it sometimes grows to a height of 150 feet, with a trunk nearly 7 feet in diameter. The timber is dense and close-grained, something like larch in colour, and like it, also, inclined to warp when exposed to strong sun-shine. It should not be cut up until thoroughly seasoned. It is strong and elastic, and well adapted for house-building purposes. It is hard, and durable when kept dry. The tree grows with great rapidity, will thrive in the poorest soils, and live in very dry situations. It makes excellent fuel, will even burn when green and wet with external moisture. The caloric powers of this wood are not excelled by those of any other kind of wood. The Fijians formerly made their best war-clubs from young *noko-noko* trees. This may account, to some extent, for large trees of this species being scarce in the group.

The *kulava* or *kukulava* (wormia biflora) grows to a height of about 30 feet, with a trunk not exceeding 6 feet in girth. It is common throughout Fiji. The wood is of a reddish colour, hard, tough, and durable. It is used for a variety of purposes by the natives and settlers.

The *savoo* or *cavoo* (Pittosporum, sp.) is a small tree, generally found on the borders of marshes near the sea. It grows to a height of 35 or 40 feet, and its trunk is seldom more than $2\frac{1}{2}$ feet in diameter. The wood is close-grained and moderately hard. It is used by the Fijians as rafters for houses, and is one of their scented woods.

Besides the above-named timber trees there are a great many others of minor importance, the value and utility of which are unknown.

Sandalwood is a well-known product of the forests of Fiji; but as information regarding it will be found in Appendix II., it is not further alluded to in this place.

Several kinds of trees and climbers, yielding caoutchouc of the finest quality, are found wild in the forests of Fiji. As Appendix I. is a report on these caoutchouc-yielding plants, further mention of them here is needless.

The Fijians require a large amount of forest produce for what may be termed domestic purposes; such as timber for house-building, wood for fuel, leaves for thatch, materials for making mats, &c.

Their houses are generally well built; and a description of one will give an idea of the kind and quantity of material required in their construction. The largest houses are those of the chiefs, the churches and schools. The length of these buildings varies from 50 to 80 feet, and their breadth from 25 to 40 feet. The average height of the side walls, from the floor to the eaves, is about 6 feet, and from the level of the side walls to the ridge about 20 feet, making a total height from floor to ridge of from 26 to 30 feet. The ridge is supported at each end by a round log,—the trunk of a tree set upright; and, should the house be long, by one or more of these logs as pillars set at equal distances apart. The tops of these logs are either forked, or hollowed out for the ridge pole to rest upon, and round logs, about a foot in diameter, are set up as side posts, at about 10 feet apart. These are also hollowed out at the top, that the wall-plate may lie solidly upon them. The wall-plates and ridges are made of the trunk of a cocoa-nut palm, or any other long straight tree; and they are not less than 6 inches in diameter. When one tree is not long enough to reach from end to end of the house, another is joined to it, by overlapping the ends, and fastening them together by sinnet. The wall-plate or ridge, as the case may be, is supported at the joint by a pillar or post. Large pieces of timber (round logs) are used as tie-beams, one at each end of the house, and should the house be long, one or more is laid across at equal distances apart. These tie-beams are fastened firmly to the wall-plates by sinnet;—binding the sides of the house together, and preventing them from being thrust out by the weight of the roof. The rafters are supported by the wall-plates and ridge, beyond which their ends extend, crossing each other and forming an angle like the letter V. The rafters are bound together by three purlins on each side of the roof, one of which is at the lower end of the rafters, outside the wall-plate, another is just below the ridge, the third midway between these two. The purlins are firmly tied by sinnet to the rafters, and so prevent any of them from shifting or sagging. The purlins are long straight pieces of timber, not less than 6 inches in diameter. The rafters are small straight trees, from which the bark has been cleanly removed, as it would harbour destructive insects. They are placed from a foot to 18 inches apart. On the outside of the above-mentioned side posts, which support the wall-plates, another row of side-posts is set up. These are generally the stems of tree-ferns smoothed, either round or square. They are placed from 3 to 6 feet apart. They are let some distance into the ground, and are fastened at the top to the lowest purlin and the ends of the rafters by sinnet. The walls or sides of the houses are formed of laminated reeds or canes of the *gasau* (eulalia japonica). The lamina are arranged either uprightly, horizontally, or diagonally. By interweaving the lamina, aided by coloured sinnet fastenings, many curious patterns are worked out. Sometimes the walls of the houses, outside the reeds, are thatched with cane leaves or branches of the *makita* (parinarium laurinum), with the leaves attached, or the fronds of the marsh acrostichum aureum. The roofs are thatched with sugar-cane leaves, ferns, or long grass. When the two latter are used, the rafters are (like the side walls) covered with several layers of reeds woven into different designs, over which the thatch is laid, all being made fast to the rafters by sinnet. When the leaves of the sugar cane are used, they are doubled over a reed and fastened, or rather sewed, by a thread-like split of bamboo, or the flexible stem of flagellaria indica. The reeds to which the cane leaves are sewed are then laid upon the rafters, beginning at the eaves, each successive layer overlapping the other more or less as the thatch, *libilibi*, is wanted to be thick or not. The site on which the house is built is raised above the level of the ground into a mound. These mounds vary in height from 1 to 6 feet, and their sides are frequently pitched with variously coloured pebbles from the rivers, to prevent the earth being carried away by heavy rains, and for ornament. After the house is finished, the floor is thickly covered with dry grass, fern leaves, &c., and then carpeted with mats. No nails are used in the construction of the houses, the different parts being securely fastened together with sinnet or the stems of strong climbers. A large quantity of sinnet is used to ornament the beams, posts, &c., and an adept, skilled in decorating and arranging designs and patterns, is much esteemed. In the V-shaped angle, formed by the ends of the rafters projecting beyond the ridge, as before-mentioned, the trunk of a long tree-fern is laid, with its ends projecting several feet beyond the ends of the roof. The projecting ends of the tree-fern are ornamented with shells, and sometimes golden coloured cowrie shells are suspended from them by a fastening of sinnet, and allowed to dangle in the air.

The houses of the common people are neither so large, so substantially built, nor so highly ornamented as those of the chiefs; but even among the poorest people there is a certain amount of pride in having their houses comfortable, and ornamented to the best of their abilities. The character of those who do not attend to these matters is much and unfavourably commented upon in Fijian society. Near the coast, and on the low-lying parts, the houses are either gable ended, or hip roofed. In the interior of Vitu Levu, the roof has the form of an ogee arch, with an outward spring at the eaves. In the middle of Vanua Levu, the sides of the houses, together with the roof, form an elliptic arch. No heavy timber is used in the construction of these houses. One end of the rafter is fixed in the ground, and the top is then joined to the top of the one opposite. These ends overlap each other, and they are then securely tied together. The rafters are kept in their places by purlins, to which they are fastened by sinnet, or the flexible stems of climbers. The purlins are laid horizontally at about 2 or 3 feet apart. The Fijians seem to be unacquainted with the use of braces or struts in the construction of houses and bridges. Their buildings are consequently kept upright by the posts being deeply sunk in the ground and firmly fixed in it.

There are several species of bamboo, natives of Fiji, and as may be supposed, they are used for a variety of purposes by the Fijians. One of these is that of a water can or pitcher. For this purpose a piece of bamboo, from 2 to 12 feet in length, and about $3\frac{1}{2}$ inches in diameter is selected. As a crack in the bamboo would render it useless, considerable care must be taken in cutting and seasoning it, and also in piercing its partitions or cell walls. One of the large sized bamboos will hold as much water as an ordinary pitcher. Like other commodities, they are articles of exchange between the inhabitants of the districts where they abound, and those residing in localities where they are scarce.

Besides in house building, the trunks of tree-ferns are largely used for making fences round the native towns to keep out pigs, which are commonly allowed to run at liberty in the woods.

The soft downy scales found on the base of the leaf-stalks or stipes of some tree-ferns are used for stuffing pillows, cushions, and mattresses. For this purpose these scales are an excellent substitute for feathers; but the softness of such articles makes them too hot to be agreeable in the warm climate of Fiji.

Although the *gasau*, a large grass or reed, may be classed among the products of waste land in Fiji, it is of great utility to the natives and settlers, in forming the side walls of their houses, temporary fences, &c. Not only are they used in forming the walls of natives and settlers' houses, but Government House and offices at Nasova are formed of triple rows of them, fastened together and bound to the framework of the building by sinnet. If not so lasting as weather-boarded houses, they are far more healthy, cool, and airy, and are altogether better adapted to the climate of Fiji, particularly when surrounded by a verandah. When kept dry the reeds will last 15 or 20 years.

The straw of the wild sugar cane, *vico*, and of the cultivated varieties, *dovu*, is of great use to both natives and settlers for thatching houses. In many districts it is indispensable, and it would be difficult to find another article, sufficiently abundant, as a substitute.

CHAPTER VII.

CONSERVANCY OF FORESTS.—DESTRUCTION BY FIRES.
—RE-FORESTING AND PRESERVATION OF WATER.

In addition to its own indigenous forest products, there cannot be a doubt that the soil and climate of Fiji will greatly favour the growth and bring to perfection the *salvian* products of other tropical countries; such as teak, ebony, sal, satinwood, logwood, mahogany, sissoo, rosewood camphor, the South American caoutchouc trees, the gutta-percha trees, &c. Nutmegs, cloves, cinnamon, camphor, candle-nuts, allspice, &c., might be grown in the forests, as forest produce, as a means of making them remunerative. These articles would scarcely pay the expense of cultivation in regular plantations; but when they can be grown in a semi-wild state they are highly remunerative. In this case the only expenses would be planting and protecting the trees, and gathering the produce,—ordinary forest expenses.

It will be incumbent on the government of Fiji to keep a considerable area of land under timber, and to plant extensively on land at present unwooded. In Viti Levu the unwooded land may be roughly stated to lie on the north-west side of a line drawn from between Serua and the mouth of Siga Toka river in the south-east, to Viti Levu bay in the north-east of the island. In Vanua Levu it lies on the north-western side of a line drawn from Udu Point in the north-east to the south-west side of Sandalwood bay, at the south-eastern extremity of the island.

From causes that will be stated, the districts in which these unwooded lands are situated are the driest in Fiji. As seen in the meteorological statement, rains are abundant in these districts at the periods of the year when northerly and westerly winds prevail in the group. At other times of the year dense mists and heavy showers are not uncommon on the tops of wooded mountains, and other places at a lower elevation, where patches of forest still remain. But these showers and fogs seldom descend to the unwooded and low-lying plains and valleys. As soon as a current of vapour bearing air passes beyond these wooded places, it either becomes rarefied by the heat rising from the unwooded land, rises to a higher altitude and passes away, or it becomes condensed and warm from parting with a large portion of vapour, and passes over the country in arid draughts until checked by a wooded hillside or patch of forest. In passing through the wooded parts of these provinces, I have frequently got a wetting, while the path through the open grassy land lying between was dry and dusty. It can scarcely be disputed that the re-wooding of tracts of land in these dry districts will be highly beneficial to agriculture. At present the rains which fall periodically flow off the land in floods in a few days. This the establishment of forests will check, and the water being entangled among the dense undergrowth, tree roots, &c., of a forest, will be absorbed by the soil, like a sponge, and parted with gradually. Further, in re-wooding the tops of the hills the soil will be kept from opening into rents by the drought, and landslips, at present a frequent occurrence in these districts, will be prevented. They take place generally at the commencement of a sudden and heavy rainfall, and are caused by the water entering the deep rents, or fissures in the soil, made by the drought, and passing downward between the soil and the stratum on which it rests, and not readily finding a way out, liquefying and carrying away the soil at the bottom of the slope. The soil composing the slope being softened and undermined at the base, and loosened from the stratum on which it rests, slips down by its own weight, and carries everything along with it.

Another benefit that may be expected to result from the re-wooding of hills, at present only covered with grass, is that the water, being hindered from running off the surface of the ground in floods, as at present, will, by percolating through the soil, be given off gradually, and a regular and abundant supply will be kept up in rivers and streams throughout the year. This is a most important consideration, not only as it will maintain a supply of water for domestic purposes, cattle, &c., but also as a motive-power for machinery, the irrigation of the land, and in navigating the rivers. Re-wooding large tracts of land will also have the effect of making the climate moist throughout the year, and, in consequence, fit for the growth of agricultural produce for which otherwise it would be quite unsuitable, by reason of aridity. For example, a field of rich land may be planted with sugar canes, and well irrigated, but if the climate of the locality be dry and arid, the crop will not pay expenses should the canes grow, which is questionable. It is my opinion that much of the cane disease, so disastrous in Mauritius of late years, is due to the unsuitability of the climate, brought about, or caused, by the over destruction of the forests. It would seem that a proportion of unwooded arable land and of forest were required to render the climate healthful for man, and to the growth of the plants which man cultivates to supply his daily wants and for the purposes of civilization; and that when the balance inclined too much either way, unhealthiness of the climate for man and his domesticated plants and animals was the result. Examples of this may be seen in various colonies, Mauritius in particular. Before the balance of forest and arable land required by the law of nature can be restored to that island, a large sum of money (about £200,000 sterling) will have to be expended in the purchase of land, planting and protecting the forests, and likely a generation will pass away before the desired results are attained. To avoid these dangers, and preserve that salubrity of climate, for which, as a tropical land, Fiji is noted among the islands of the Pacific, it will be necessary for the Government of the colony, in the disposition of lands, to set apart large forest reserves in both the wet and dry districts. These reserves should include, if possible, all the mountain and hill tops (in a proportion of one-third of their slopes), the land surrounding springs, and at the watershed and sources of rivers and streams; in short, the timber should be preserved on the land on which the WATER in the streamlets, &c., *is collected*. To effect this properly, and perhaps for other reasons, it would be unwise to dispose of more than one half of the land at one time in any district, even after the forest reserves have been set apart. After the lands disposed of have been settled and cultivated, and the country opened up by roads, &c., the remaining half of the lands will be easily and profitably leased or sold. And as some climatical experience of the district will have been obtained, it will be known to a certainty what farther extent of land may be given up to cultivation without injuring the climate or fertility of the soil. In addition to its beneficial effect upon the climate, soil, &c., there are other advantages of great value to a community to be derived from keeping large tracts of land under forest, viz., an abundant supply of cheap fuel, timber for house and ship building, for sugar factories, and all other industrial purposes connected with the colony. In this way large sums of money will be kept in the colony that otherwise would have to be sent out of it annually. The case of Mauritius may be quoted as an example of this. From that small, though not unimportant island, about £20,000 is sent annually to India, Singapore, and Australia for timber, the produce of tropical countries, which might have been grown in the colony. This sum does not include the value of pine imported from Europe, which could not be grown in Mauritius. The Fijian *dakua* and *dakua-salu-salu*, are equal if not superior to any pine for all the purposes for which it is used, and quite as easily worked. It may therefore be anticipated that by the sale of the timber, and other forest produce, these forest reserves will not only be self-supporting, but when the colony is fully settled, and the demand for various forest products greater than at present, they will yield an annual revenue to the colony. There are persons who hold the opinion that no timber ought to be felled in Government reserves, which they say are maintained solely for climatic purposes. This is equivalent to taxing the community to keep up property from which they receive only half the benefit which that property could yield. By the sale of mature timber, these reserves (as has been stated) will in time be more than self-supporting, without in the least injuring them for those purposes for which they were established. The objects to be obtained in this matter are of so great importance to the colony, that Government cannot trust the carrying out of them to private persons, who will study their own interests rather than the welfare of the commonwealth. The vital importance of the subject is my only excuse for dwelling so long upon it. Of course results commensurate to the foregoing ideas will not be realised in a year, nor perhaps for some time to come; but if the object to be attained is kept steadily in view by the Government, and intelligently worked up to, there is not much fear of ultimately attaining a satisfactory result. It may be added that the future success of the matter depends very much on steps being taken at once to lay a solid foundation upon which the structure is to be built.

Reference has already been made to the unwooding of the leeward districts by the system of cultivation practised by the Fijians, and destructive fires which annually sweep over these districts in the dry season. The mischief caused by the former of these in all parts of the group has been alluded to, and I would now call attention to the latter. These fires originate in several ways, either when burning the scrub and grass in clearing land for a new plantation and carelessly allowing the fire to spread into the surrounding dry herbage; (2) by sparks from torches carried by the natives when travelling after nightfall to show the paths and fords of rivers; (3) or by lighting fires to cook food and dry their smoking tobacco when on a journey, as is the custom of the Fijians, and leaving them unextinguished. With reference to the first of these, it is either from want of forethought, carelessness, or perhaps indolence, that fires kindled to consume rubbish, are allowed to spread beyond the boundary of the clearing. Such carelessness ought to be severely punished. As to the second, good paths and roads and bridges will, in a great measure, do away with the necessity of carrying torches. Still, naked lights ought not to be allowed among dry herbage, at least during the dry season. In respect to the third, ground in the vicinity of the resting-places should be cleared of all herbage, for some distance round, and any party kindling a fire should extinguish it before leaving, under a penalty. A further precaution would be to surround each reserve by a fire-break of at least 12 yards in width, from which all inflammable material ought to be cleared away annually. In addition to this, the roads between the blocks or divisions of the forest should be kept clear of everything that will burn, so that, should a fire occur in one block, it may be prevented from spreading into others. When the colony is fully settled, and the open grassy lands in the dry districts pastured by cattle or sheep, which will eat down the long grass at present covering these lands, the dangers from fire will be much lessened, especially if regulations against fires are firmly enforced.

It is not to vegetation alone that these fires have done injury. By burning the grass off the ground the soil is laid bare, and the surface is cracked and crumbled into dust by the heat of the sun. From the steepness of the ground in some places, the first heavy rain carries off the loose surface soil to a depth of several inches. The frequent recurrence of this has resulted in laying bare the subsoil in some parts of the country.

In some districts, these tracts are now covered with hardy ferns and rough kinds of grasses; but in other places of small extent, the subsoil lies exposed on the surface. Unfortunately these fires are still occurring, and only last dry season they laid waste a great part of the province of Navosa.

The road (or rather bridle path) from Nadi on the west coast of Viti Levu, to Fort Carnarvon at na Tua-tua-coka in the interior, passes, on the top of the mountains, through the charred remains of what had been a large and magnificent forest of the finest timber trees in Fiji, now completely ruined by fire. This happened about six or seven years ago. When I passed through the forest in August 1878, the dead blackened trunks, with leafless branches, were still standing like great chimney stalks, but the undergrowth had been completely burned up. The natives from the neighbouring towns had taken advantage of the circumstance and planted large patches of the ground thus cleared with yams, *dalo*, and bananas. In some places a natural growth of young trees and shrubs had started, but most likely it has been destroyed in the conflagration of last year, which occurred after my departure from Fiji.

What may be termed the natural tendency of the land to re-wood itself should be taken advantage of; and there can be no doubt that large tracts of land on the leeward side of Vitu Levu and Vanua Levu, would speedily be covered with forest of a natural growth, were the ravages of fire prevented, and browsing animals, as cattle, horses, sheep, goats, &c., excluded.

Before any tract of land is set apart to allow it to become naturally re-wooded, it should be carefully examined, and all symptoms of natural growth noted. This will prevent money being expended in protecting land on which a natural growth of trees might not spring up. It would be incorrect to affirm that most of the forest trees thus naturally produced would be of utility to the builder; still, natural production would be a ready means of serving a climatic purpose, attended with little expense, and would give time for the more valuable kinds of timber being planted in other parts of the colony. In the meantime planting should be confined to places which will not become re-wooded naturally. The kinds of trees selected for planting should not only be adapted to the climate and soil of the locality and district, but should be chosen with an eye to their future value, whether for fuel or timber; as caoutchouc or sandalwood yielding, or spice plants.

By carefully selecting the most valuable forest trees indigenous to Fiji and other tropical countries, a large number of useful kinds could be obtained adapted to the climate and soil of the various districts of the colony. A few exotic kinds may be mentioned. For the poorest soils of the dry districts, hæmatoxylon campeachianum, logwood; albizzias lebbek, odoratissima, elata, moluccana; tetranthera laurifolia; acacias arabica, eburneum, and catechu; ingas dulcis, and xyclocarpa; adenanthera pavonina. Among the kinds of timber trees that will succeed in localities that are neither particularly wet nor dry may be mentioned teak (tectonia grandis); sal (shorea robusta); hopeas maranti and stipulosa; vaterias indica, ceylanica, and seychellarum; dalbergias sisso (sissoo), and latifolium (rosewood); pterocarpuses indicus (santal), and marsupium (kino); cedrela toona (toon); lagerstrœmias parviflora, and regina; bassias latifolia, and longifolia; swietenia mahogani (mahogany), swietenia chloroxylon (satinwood); diospyroses ebenum, and hexandra (both yield ebony); isonandra gutta (gutta percha tree); ficus indica, and the caoutchouc yielding castilloas, heveas, and manihot Glazovi of South Africa. It will ultimately be found that some of these will succeed in the wettest parts of Fiji as well as in moderately dry places. Although some of these trees may grow in the dry districts, yet judging from their nature, planting them in dry places cannot be recommended in a general way. A good deal of re-wooding could be done inexpensively, by sowing the seed *in situ* in the manner indicated for the caoutchouc and sandalwood trees (see reports on these subjects, Appendices I. and II.). However, it is only the indigenous trees that produce seeds abundantly, that can be profusely increased; exotic trees will not be easily extended until they produce seeds freely in the colony, and the usual method of planting, where it may be desirable to establish them, will have to be resorted to for some time.

A committee of Woods and Forests was appointed by His Excellency the Governor of Fiji to confer on forest matters, and several meetings and consultations were held, but from my almost constant absence from Levuka, the subject was far from being exhausted when I left Fiji. Before my departure, I drew up propositions for a Forest Ordinance, and gave them to the Colonial Secretary for submission to the Government. Since I came to England I have arranged these propositions under proper headings, and they are printed in Appendix III., with explanatory remarks or reasons for each proposition, printed on the opposite page. It may be stated in reference to the forest staff, that a large establishment is not required, at least in the meantime; but it is necessary that a person with a fair knowledge of forestry be appointed to regulate and superintend the felling of timber on crown lands, marking off reserves, &c. See further remarks on this subject, under suggestions for Botanic Gardens. It need scarcely be added that subordinates should be taken on as required, and that the "Forest Department" should increase in importance along with the colony.

CHAPTER VIII.

BOTANIC GARDEN.—INDUSTRIAL SCHOOL.—MUSEUM.
—METEOROLOGY.

In a new colony like Fiji, a Botanic Garden would be of the greatest importance and utility, independently of the knowledge of botany as a science, and the taste for plant culture which it would foster and diffuse in the colony. It would be highly popular with the settlers, and the subject has frequently been mentioned with approval at the meetings of the Fiji Agricultural Association. No information has yet been obtained as to what action the Government of the colony have taken on the subject.

The introduction of many useful plants into Fiji has been suggested, and the desirableness of having a suitable place for planting them need scarcely be alluded to. In it the plants would be properly grown and propagated, and distributed from it to the inhabitants in all parts of the colony. From practical observation the settlers and natives would become familiar with the treatment or field cultivation which different kinds of plants required, and the preparation of the produce for the market. It would be the duty of the Director or Manager to spread throughout the colony all the information he possessed or could obtain from the best sources as to the nature of the plants, kind of soil, and treatment or culture which they required. Reliable practical information on tropical cultivation is much needed in Fiji. Most of the settlers know next to nothing about the husbandry of the sugar cane, coffee, tea, cinchona, cacoa, and other products of the tropics. The consequence is that large sums are spent on worthless experiments, or in working out useless theories, and not unfrequently on fanciful and absurd notions. This want of practical knowledge among the settlers has been the ruin of Fiji from an agricultural point of view. Nor is the Government of the colony, in agricultural and foresting matters, in a better position than the settlers. All are groping in the dark, and the want of a living reliable authority, to whom all could refer and obtain good practical information, is very much felt. Indeed the date of supplying this want will mark a new era in the agricultural condition of the colony, since a supply of labour to work the fields has been secured, and can now be depended upon.

A Botanic Garden might be established on a small scale, as a nursery at first, and be extended and beautified as the colony increased in prosperity, and became able to support one, such as that in Ceylon, Mauritius, Trinidad, &c. This ultimate increase should be kept in view from the beginning, and the means of attaining it, proportionate to the resources and needs of the colony, carefully considered, laid down, and intelligently worked up to. A considerable area, about 30 acres, would ultimately be required as nursery ground, on which to rear young plants for distribution to the inhabitants, and for planting in the forests. Besides this, space for the growth of fruit, spice, timber, and caoutchouc trees, &c., would require to be provided, in order that seeds might be obtained in quantity sufficient to meet all future demands. Park-like space would best answer this purpose, and it would most easily be kept. In addition to these, it might be desirable to have within the same boundaries of land room for the acclimatisation of useful animals, and a model farm for the trial and acclimatisation of different kinds of agricultural plants, such for instance, as rice, different kinds of coffee, sugar cane, cacoa, tea, cinchonas, gram, and various other grains. About 50 acres would be required as Botanic Gardens and Pleasure-grounds. Altogether about 300 acres might be set apart for the purposes mentioned. At the commencement, at least, the Director or Curator of the Botanic Gardens might take charge of the forests. Not that there would not be work enough for him without them, but because supporting two highly skilled men might be too heavy for the resources of the new colony.

The Botanical Gardens should, if possible, be within an easy distance of Suva, the chosen site of the Capital, and also near the highway into the interior of Viti Levu. A site in the vicinity of the native town of na Colo-suva, lying between Suva and the Wai Manu, would answer the purpose admirably, as it would combine a suitable climate, good soil and a supply of water, being within 5 miles or so of the capital, with access to the interior of Viti Levu and all the other islands of the group by the eastern branch of the Rewa river.

It may be suggested that the residence of the Governor might be in connexion with the proposed Botanic Gardens or adjacent to them. This would do away with the necessity of keeping separate grounds.

The founding of an Industrial School in Fiji well merits consideration. For this purpose advantage might be taken of the present training school at Navaloa, Viti Levu. To that institution the more advanced and most promising young men are drawn from the native schools which the missionaries have established in all the towns throughout Fiji. They not only receive a good education, but are employed a part of the day, before or after school hours, at field work, cultivating yams, dalo, &c., on the land belonging to the institution. In this way the institution is altogether, or to a great extent, supplied with food, and the exercise is most beneficial to the health of the scholars. The system of agriculture practised by the Fijians has been mentioned, and the evils arising from it pointed out. As it may be desirable to change this system for one which would not require new ground every year on which to grow food plants, advantage might be taken of the institution at Navaloa to impart to the natives a thorough knowledge of modern agriculture. There the young men would see the result of using manure; that by its use the same kind of crop can be taken repeatedly from the same ground, and also the benefit of a yearly rotation of different kinds of crops. Not only so, but as a part of their education, they would become acquainted with the various practical ways in which these results are brought about.

The young men trained at this institution, the most of whom are the sons of chiefs, will naturally become prominent and influential members of the native community, and it may be expected that the knowledge thus imparted to them will soon be extended to others. Besides this, they might also be taught the use of tools, how to work a saw, drive a nail, use a plane, and handle an adze, &c.,—a kind of knowledge they much need, and which would be most useful to them in their after life.

At Levuka, the present capital of Fiji, there is a Mechanics Institute, and a fairly good library, circumstances considered. The Government having granted a site for a new building, to replace the old dilapidated and incommodious one at present occupied, advantage of the circumstance might be taken to form the nucleus of a Museum for the colony. The Society which the institute represents is neither numerous nor wealthy; but it would doubtless lend its aid in this matter for the information of the people. Towards this end Government might help the Society by granting a small sum annually upon well considered and approved conditions, say of giving accommodation in the building for a small but increasing collection. This collection might represent the flora of the colony; its land and sea fauna, mineralogy, geology, and industries; and articles of curiosity, and of historical and economical, commercial, or educational interest from other countries. The co-operation of naturalists visiting the colony should be desired, and they might be requested to give the museum one specimen of each species of the collection made in the group. The settlers and natives in all parts of the colony might contribute, and a fairly representative and interesting collection might be easily and rapidly got up. The habits of the natives are changing, surely, if not rapidly, and articles of interest such as represent their former condition, manners, customs, clothing, manufactures, &c., are fast disappearing, and ere long there will be more of these articles in Europe and Australia, where, as curiosities, they are more valued than in Fiji. A Museum would retain a fair collection of those articles, which it would be desirable and interesting to preserve in the colony.

The climate of Fiji is essentially tropical, yet very healthy. Malarial fevers and other diseases common to nearly all tropical countries are entirely unknown in Fiji. New comers, however, are said to be subject to attacks of diarrhœa or dysentery. This may arise from careless living, or change of diet; or, if the person has been travelling in the colony, from poorness of food and not being acclimatized or accustomed to having the clothes wet, and neglecting to change them. Slight wounds have a tendency to become sores, if not at once properly attended to, kept clean, covered from the air and strong sunshine, and from salt and brackish water. The Fijians are certainly a robust, healthy race, and live to a great age. Though in some places their houses are situated on low-lying marshy land, or even in the middle of marshes of brackish water, and on the edges of mangrove swamps, the people are exempt from all complaints arising from miasma, beyond an occasional severe cold in the cool season, asthma, or rheumatics.

Elephantiasis is common among the natives, and so is an ulcerous disease locally called *coko*. It is not fatal, and is mostly confined to children. The Fijians say that their children are neither strong nor healthy till they have have had it. It is supposed to be hereditary, but not contagious. A kind of ophthalmia, which lasts only a few days, is not uncommon, and both natives and settlers are subject to it. The settlers throughout the colony, and European tradesmen in Levuka, such as carpenters, &c., work in the open air constantly throughout the year, exposed to all the changes of weather, without feeling any bad effects. This says a great deal for the healthiness of the climate, as Levuka is about the worst place in Fiji for situation. But however healthy the climate of Fiji may be in general, its effect on the average European is enervating and depressing. This is particularly the case during the hot days that occur in the months of December, January, and February. A heavy, languid, oppressive feeling is experienced, accompanied by an unwillingness for the least exertion, either mental or physical. The relaxing effect of the climate renders a change to a cooler region at times desirable, if not necessary. The dull indolent habits of the natives, too, have a depressing effect on those who are much in contact with them, and it needs the quickening influence of mingling with a superior race to sharpen the thoughts, as well as the cool air of a temperate climate to brace and invigorate the body.

In Fiji, as in most tropical countries, there is a dry and a wet season; the former is cool, and lasts from May to October, the latter is hot, and lasts from October to May. In the dry season, the south-east trade winds prevail, and every person is benefited by the cool invigorating breeze. During the wet season there are frequent calms, and the winds are variable, though generally from a northerly direction. The weather is hot, and the least exertion brings the perspiration in streams out of the body. While the cool weather lasts, Europeans can wear, with comfort, clothing adapted to an English summer; indeed, at this season the weather is delightful,—finer than the best summer weather in England. In the hot season the least amount of clothing is burdensome and oppressive. Thunderstorms accompanied by vivid and rapid flashes of lightning are common all over the colony during the hot season. In the cool season these storms are of rare occurrence, and the wind blows steadily, and frequently in strong breezes during September. From the beginning of December to the end of March, owing to the backing and filling of the north-west monsoon, the winds are unsettled. At this time of the year, both extremes meet, and calms are followed by sudden squalls. These squalls are accompanied by deluges of rain, and may last from half an hour to several days, during which time the wind will blow from all points of the compass. But although a year seldom passes without one of these circular gales occurring in one or other of the islands, they rarely if ever rise to the fury of a hurricane, such as those that annually sweep over Mauritius or the West Indies. The houses of the natives, the settlers' buildings, trees, &c., are evidence that heavy gales of wind (hurricanes) do not occur.

The annual rainfall in Fiji may be considered heavy even for a tropical country; and what is of the greatest importance in an agricultural point of view, is that the rains fall most abundantly during the warm or summer season, when vegetation most requires it. As tables of rainfall, &c. will be found in the App. (V.) it is unnecessary to go into details; but it may be interesting to state that Delanasau, on the north coast of Vanua Levu, is one of the driest parts of Fiji. There, in 1877, rain fell more or less on 106 days; at Qara Walu, in Taviuni, on 228 days; at Levuka, Ovalau, on 158 days; and at Wai ni Sasa, on the Rowa river, Viti Levu, on 139 days. Unfortunately we do not have the quantity that fell at the last-mentioned place. Wai ni Sasa is on the banks of the Rewa river, about 25 miles from its mouth. The district is popularly known as one of the wettest in Fiji; but as far as a comparison can be made, the number of days on which rain falls is less there than at Levuka. At Delanasau, in 1877, December was the hottest month of the year. the mean temperature being 84·1 (max. 97·6 ; min. 70·8), see table I. At Levuka, in the same year, February was the hottest month, the mean temperature being 82·2 (max. 92·2 ; min. 73·2), see Table II. The first four months of the year, and the last two, were the warmest at Delanasau, Table III. Table I. shows the mean annual rainfall, temperature, &c., at Delanasau for seven years ; but as there are no observations for the same number of years from any other part of the colony, a correct comparison cannot be made. An important fact, indicated by Table II., may be noticed. Previous to, and during 1861–2, the low hills around Levuka were thickly wooded. Since that time the woods have been cut down, and the number of days on which rain falls has been reduced from 256, the average for 1861–2, to 149, the average for 1865–6 and 1876–7. It would seem that the number of showers diminished simultaneously with the cutting of the trees. The thick woods afforded shelter to the mountaineers, who on several occasions appeared in large numbers, and threatened to sack the town and murder the white settlers. These marauders came from Lavoni, in the centre of Ovalau, just across the mountains from Levuka ; stole down upon the town ; plundered the goods of the settlers, and then made off into the woods, where it was useless and dangerous to follow them. In consequence of this, the woods in the vicinity of the town were entirely cut down. But it may be remarked, that the average rainfall for the year does not seem to be much affected by the cutting down of the woods. In answer to this, it may be stated that whereas formerly the rain came in gentle showers which sank into the ground and refreshed the vegetation, it now descends in torrents, and runs off the ground, carrying away the loose soil on the surface, where the ground is steep, and doing great damage to both soil and vegetation.

Table VI. shows that from observation taken during five years, the last three months of the year, with the first four months of the following year, constitute the wet season at Delanasau, and doubtless, for Fiji. These tables, however, give merely an indication of the climate of Fiji. The islands differ a great deal as to the amount of rainfall; and the temperature of different localities varies so much, that a correct idea cannot be given until observations have been taken at many different parts of the colony and compared.

To judge from the number and size of the rivers, and the volume of water which they pour into the sea, combined with the small area of the river basins, it may be concluded that the yearly rainfall in many parts of Viti Levu and Vanua Levu is greater than that registered during two years at Qara Walu, in Taviuni (See Table IV.). At some places on the north coasts of these two islands the rainfall may be less than that so carefully recorded at Dalanasau.

As regards temparature, an observation can hardly be said to do more than indicate that of the locality where it was taken. In the interior of both the large islands, the ruling temperature is very much lower than that noted on any part of the coast. Again, on the elevated plains, or grass-covered hills in the interior of Viti Levu, the temperature during the day is sometimes as high, if not higher, than on the coast, while during the night it is much cooler,—perhaps cold for a tropical country.

Observations should be taken in a systematic manner and at stated hours of the day. Approved printed forms, ruled and divided into columns under the required headings, as in Table II., Appendix V., might be issued for each month of the year, to the observer at each station. Failing a central observatory, monthly returns of observations might be sent to the Registrar-General of the colony, to be compared and published yearly or monthly in the Government Gazette. These forms could be printed at small expense. They would save a great deal of time and labour to the observers in monthly ruling forms and dividing them into columns, copying headings, and columns of figures. The observations could be put down on the form at the time they were taken. The Registrar-General's address should be printed on the back of the forms, and they might be returned to him free through the post office.

Observations might be taken by the police at their several stations throughout the colony, and at the court-houses in the different islands and provinces. The aid of the planters might be invoked in this measure, so important to agriculturists. The study of the weather is as important to the planter as it is to the seaman; and observing the thermometer, barometer, rainfall, &c., and keeping notes and observations on them, is the way to attain a correct knowledge of it. Properly used, such knowledge saves much expense and unprofitable labour to the planter in the tropics.

CHAPTER IX.

NUMBER OF ISLANDS IN THE GROUP—AREA OF LAND—REEFS—RIVERS—NAVIGATION—WATERSHEDS—MOUNTAINS—LAKES—TAVIUNI—RABI—LOMA LOMA—KORO—OVALAU—LEVUKA.

The number of islands comprising the Fiji group is over 200, and their aggregate area is estimated at about 7,400 square miles. Viti Levu, the largest island, has an area of about 4,100 square miles, Vanua Levu 2,432, Taviuni 217, and Kadavu 124 square miles. The other islands are of small extent, from Koro, 57 square miles, downward.

Most of them are surrounded with reefs of coral. These reefs are of two kinds, "barrier" and "fringing" reefs. Frequently both kinds of reefs may be seen at the same island, and at one part of its coast. At another part, the reef may be a barrier one, and again at another place a fringing reef. There are numerous deep-water channels through the barrier reefs, and on the inside of these reefs there are several fine harbours, notably those at Suva, Levuka, Savu-savu, &c. There are commodious roadsteads in the group, where ships of the largest size may anchor in safety, and leave with almost any wind. Between the islands, and inside the barrier, there are numerous patches or "heads" of coral, dangerous to strangers, in navigating among them. Still these reefs break the swell of the ocean, and render navigation throughout the group safe, even for open boats in ordinary weather.

The islands lie in small groups of several together. In some places the water is so deep that vessels of 600 tons can be moored to the shore; while in other parts of the same island, the water is so shallow that a ship cannot get within several miles of the shore.

Most of the islands are high and mountainous, rising abruptly from the sea as if they were the mountain tops of a submerged continent, or large island. The mountains of Taviuni, Vanua Levu, Viti Levu, and Kadavu rise to an elevation of 3,000 feet above the level of the sea. In none of the other islands do they much exceed a height of 2,000 feet. There is very little level land in any part of Fiji. Hills and lovely valleys follow in rapid succession, from the shore to the interior. Lofty peaks rise in a succession of precipitous ascents, one after another, sending off numerous spurs, which again subdivide and ramify in a manner that defies description.

In many instances the sources of streams overlap each other, and the waters from parallel valleys find their way to the sea in opposite directions. Near Namosi, a stone, dropped from where the Wai-Dina rises, would fall into the waters of one of the principal branches of the Navua river, which at this place is a rapid stream 30 feet broad and 2 feet deep.

At a place near Nadrau, a similar thing occurs, viz., from where the Wai ni Mala rises a stone could be thrown into the Siga Toka river. In Lavoni valley, Ovalau, two large streams from opposite directions unite and flow off at a right angle to their former course. Many more instances might be given, but these will suffice to show the complications of the watersheds in Fiji.

In Vanua Levu a range of lofty mountains extends almost from end to end of the island, and forms its main watershed. In the mountain and river system of Vanua Levu, the principal slope is towards the west and north-west. The chief subdivisions of this watershed are in order from south-west to north-east; (1) in the vicinity of Dand's Peak, (Corobato); (2) The Sugar Loaf (Thorobala); (3) Needle Peak; (4) Drayton's Peak; (5) Hale's Peak. From the first, the water flows eastward towards Kobalau Point and Savu savu Bay, and north-westward towards Naloa. Bua is surrounded by a semicircle of mountains, among which, in a valley, flows a large stream that rises in the vicinity of Dand's Peak, and enters the sea a little to the south of Sandalwood Bay. From the Sugar Loaf (2), the water flows eastward to Savu-savu Bay, and south to Wai Nunu near Kobalau point.

The tributaries of the Dreketi river rise in the vicinity of Drayton's Peak (4), and, flowing in a westerly direction, join the main branch of the river as it flows from the mountains (Needle Peak) of Wai Levu, falling into the sea nearly opposite the island of Nuvera. It is navigable for small craft drawing 3 feet of water for 25 miles from its mouth. A range of mountains extends along the N.W. coast of Vanua Levu, from near the mouth of this river, to nearly opposite Mali island. Between this range and the main watershed, which runs in a zig-zag manner through the island, lies the basin of the Dreketi river and its tributaries, four large streams. From Drayton's Peak, the watershed runs eastward, close to the N.W. shore of Natawa bay, and northward to Hale's Peak (5). The waters from this basin flow in a north-westerly direction, and enter the sea by several large streams opposite Mali island. Again, on the north of Hale's Peak another river basin is formed, bounded on the south by Hale's Peak, on the south-east and east by the main watershed, which here runs close to the shore of Natawa bay, and on the north-east and north by high land, which crosses the island from Tibethe point. The waters collected in this basin flow also in a north-westerly and westerly direction, and are discharged opposite Sau-sau passage near Vuni-vuti, and Tutu, near Drua-drua island. Several streams fall into Natawa bay on its N.W. and S.E. coasts, as at Koro-ni-saca, Malaka, and Wai Levu at the head of the bay.

These streams are all navigable for boats and small craft, for some distance into the interior, notably those which fall into the sea at Mali island and at Tutu. From this place I went up the river 8 miles in a canoe loaded with ten people and half a ton of luggage.

No streams worthy of mention enter the sea on the long line of coast from Savu-savu bay to Somo-somo straits.

The configuration of the land is the cause of the N.W. (Macuata) coast of Vanua Levu suffering from drought, except during the northerly winds. The range of mountains forming the watershed, and extending in a S.W. and N.E. direction, are, as has been stated, covered with dense forest from end to end of the island. On these mountains, the vapour laden clouds, carried by the S.E. trade winds, settle, and pour out rain in torrents; and the air having parted with its moisture passes over Macuata in dry gusts.

Without speaking of the peaks mentioned, which are much higher, this range of mountains rises in many places to a height of 2,000 feet above the sea. The estimated heights over which I passed, in crossing and recrossing this range, are as follows:—From Koro-i-vono to Togaloa, at a height of 1,500 feet. From Vuni Sawani to Savu-savu at 1,000 feet. From Wai-wai, Savu-savu bay to Naduri at 1,800 feet; Nadoga to Malaka at 1,500 feet; and from Bua to Wai Nunu at 1,000 feet.

In passing over these mountains many striking and frequently magnificent views present themselves. Here forest and woodland, with valley opening into valley in oft repeated succession; there on one side the open, undulating grass-covered country of Macuata; on the other, the blue sea studded with islands —with spots and lines of white foam where the sea is breaking on the reefs; all these, seen from a considerable elevation, combine in forming a panorama of which words can convey only a faint idea.

Natawa bay, in the N.E. of the island, runs about 30 miles inland. At its S.W. extremity the distance to Savu-savu bay is about 7 miles, and about $2\frac{1}{2}$ miles to the south coast. At one place this narrow neck of land, joining the peninsula thus formed to the mainland, is not more than 100 feet above the sea.

At the narrowest part there is a salt lake of about 150 acres in area. It is surrounded by hills, and has communication with the sea by a deep narrow channel, about 6 yards across at its narrowest, and about a mile in length. Through this channel the tide runs in and out with great velocity. From the appearance of the flat land which borders the lake, it is conjectured that the lake was at one time nearly double its present size; and the narrow ridge of land which separates the lake from Natawa bay, also bears unmistakeable evidence of having been under water. The lake and the channel to the sea are bordered with the kinds of plants usually found on the sea-shore, and the rocks in its vicinity are of that breccia so common in Fiji. Coral is said to abound in the lake. It is most probably the crater of an extinct volcano. From the lake, the Fijians drag their canoes across the narrow ridge of land, and descend with them into Natawa bay.

Lakes are not numerous in Fiji. There is one in Taviuni at about 3,000 feet above the sea, in an old crater. Near the native town of na Kali-kosa, in the interior of Vanua Levu, there is a small lake with a floating island in it.

On the south coast of Vanua Levu, near Mr. Hawkins' cocoa-nut plantation between Levuka and Vatu Kali, a singular occurrence has taken place. At the place in question, the level of the reef falls suddenly about $2\frac{1}{2}$ feet, like a step. This fault runs right out to sea at right angles to the coast-line. East of this point there is a double line of reefs, an outer or barrier reef, and an inner or fringing reef. To the westward there is only the fringing reef. This would lead to the conclusion, that there has been either a depression on one side, or an upheaval on the other. In connexion with the whole character of the coast to the west of this point, and the indications of the salt lake having at one time covered a larger area than at present, an upheaval is most probably the correct inference. The surface of the flat land bordering the lake is from $2\frac{1}{2}$ to 3 feet above the level of the water.

Rivers of Viti Levu.—Visitors to Fiji, coming from the "Colonies," are greatly surprised at the size and beauty of its rivers; and, considering the size of the islands, the number of the rivers and their magnitude are certainly astonishing. The largest river in Viti Levu is the Rewa—the *Wai Levu* (Great River) of the Fijians. It is formed by the junction of the Wai ni Buka and the Wai ni Mala, and afterwards receives on the right bank the Wai Dina and the Wai Manu. At 25 miles from its mouth it is about 200 yards in breadth, and at its ordinary height has a volume of water nearly equal to that of the Tay at Perth.

The Navua river is the next in size, up which vessels drawing over 6 feet of water can go for 15 miles. The next is the Siga Toka, which is perhaps the longest river in Fiji. Unfortunately there is a bar at the mouth of this river which makes it dangerous to enter from the sea, but inside there is deep water. The Ba river is the next in size. It is also navigable for several miles from the sea. Then follows the Wai Delice, up which boats with about 10 tons of cargo can go for 17 miles. Besides these, there are a great number of smaller rivers, the most of which are navigable for several miles inland.

Branches of the Rewa.—The Wai ni Buka rises in the mountains that border the northern or Raki-Raki coast of Viti Levu, on the western side of Viti Levu bay, and flows in a south-easterly direction to its junction with the Wai ni Mala. In its entire length it runs parallel to the eastern coast, from which it is separated by a range of high wooded mountains. The Wai ni Soga, one of the branches of the Wai ni Mala, rises in the same mountain range as the Wai ni Buka, and flows in a south-easterly direction to na Babuca, where it is joined by the Wai ni Loa, or Black water. This river rises on the high table land, overhanging the banks of the Siga Toka, near Nadrau, and flows in an easterly direction to na Babuca where it receives the Wai ni Soga, then runs in a south-easterly direction and joins the Wai ni Mala, above Koro Suli. The Wai ni Mala rises in the mountains about na Qara-wai, and flows in a north-easterly direction to its junction with the Wai ni Loa. The united rivers under the name of the Wai ni Mala flow eastward in very zig-zag courses to a little below na Koro Vatu, where their waters unite with the Wai ni Buka in forming the Rewa.

A branch of the Wai-dina also rises in the vicinity of na Qara-wai, and flowing in a crooked south-easterly course, joins the Wai-dina at na Bukè Lukè. The Wai-dina rises about 4 miles west of Namosi, at a place where a stone could be dropped into the waters of a branch of the Navua, and flows east by north in a very winding course to the Rewa. The Wai Manu rises on the eastern side of a mountain called Kora Loa, not far from the Navua river, flows in an easterly course nearly parallel with the Wai Dina, and falls into the Rewa about 20 miles above its mouth.

This river system is a most important one. The Wai ni Mala is navigated by canoes drawing $2\frac{1}{2}$ feet of water to Koro Suli, a distance of about 30 miles by the river, above its junction with the Wai ni Buka. The Wai ni Buka is also navigated for about an equal distance. Light draught canoes go up the Wai-dina 20 miles above its junction with the Rewa, and barges with 20 tons of sugar canes come down the Wai Manu a distance of about 8 miles above its mouth at na

Vusa. By the Wai ni Mala and Wai ni Buka, a system of river navigation is continued down the Rewa to the sea, a distance of 85 or 90 miles by the course of the rivers. Flat-bottomed boats containing about 15 tons of cargo, are brought from Messrs. Scott and Harvey's settlement at the junction of these two rivers, down to the sea, a distance of 55 to 60 miles.

The Navua River.—A branch of the Navua river rises in the mountains near na Qara-wai, and joins the main stream at Bega. The main stream is composed of several large tributaries, which have their sources near the banks of the Siga Toka, and on the inland slope of a range of mountains which commences at Serua, and runs in a more or less broken range to the Rewa river. The Navua is a rapid river, and its navigation is dangerous in consequence of the Namata falls, about 25 miles from its mouth; but the natives convey produce from the vicinity of Namosi, and above Bega on the other branch, down to the sea. For this purpose, rafts constructed of bamboos are used, and as the river is a series of deep pools, rapids, and sudden turns, it requires great skill and nerve to steer either a raft or a canoe safely to the sea

The Siga Toka.—The main branch of the Siga Toka rises near the sources of the Wai ni Buka, in the mountains near Raki-Raki, passes Nadrau, and traverses Viti Levu from north-east to south-west. On the left bank it receives the Wai Suli Kana about 10 miles above na Tua-tua-coka, or Fort Carnarvon. This branch rises on the western slope of the mountains near na Qara-wai, and flows in a westerly direction until it falls into the main stream. A third, but much smaller branch, rises on the eastern side of

Pickering's Peak (Koroba), flows east by south, and joins the main river at na Tua-tua-coka. The Siga Toka is perhaps the longest river in Fiji; but its course lies at the bottom of a deep valley, which narrows in some places to a perfect gorge or chasm, at the top of which, and almost within a stone-cast of its channel, some other rivers (notably the Wai ni Mala, &c.,), have their sources.

The Ba river is formed of two branches, one of which rises in the mountains behind Raki-Raki and flows for some distance in a south-westerly direction, then turns and flows in a north-westerly direction to the sea. The other branch rises on the top of the gorge overlooking Nadrau on the Siga Toka, flows first to north-west, then turns suddenly to north-east, and joins the main stream about 25 miles from its mouth.

The Wai Delice rises in a ridge of mountains which stretches along the coast with diminishing elevation, from Namena to Viti Levu bay. The course of the river, at first is in a southerly direction, then taking a turn to the eastward, it reaches the sea near Verata point. *The Nadi river*, which is of second or third rate importance and size, rises between the right or main branch of the Siga Toka and the left branch of the Ba river, and flows westward into Raurau bay. On its left it receives several small rivulets from the northern slope of Koroba mountain.

Mountains and main watershed.—The main watershed of Veti Levu begins on the south coast near Serua, and runs almost due north through the centre of the island to the Nananu islands at its most northerly point. The most important slopes are towards the south-east and south-west, only one river of any size having a north-westerly course, viz., the Ba river. The slopes on the south-east side of the main watershed are well wooded, and receive the south-east trade wind directly from the ocean. On them the clouds of vapour brought by this wind, break, and pour out rain like a deluge. The country on the N.W. side of the watershed is denuded of trees, excepting a few small patches of forest on the summits of some of the mountains and on the steep rocky sides of a few ravines. The ascending land and mountains on the S.E. side of the watershed are the chief cause of the descending slopes and hills on the N.W. side getting little rain, during the time (6 or 7 months each year) that the S.E. winds prevail in the group. Also, the generally unwooded state of the country on the N.W. side largely contributes to the same effect. On the other hand, when north-westerly winds predominate the N.W. side is the wettest, at least its high-lying lands and its best wooded parts are so.

Voma, in the vicinity of Namosi, is supposed to be the highest mountain in these islands; but there are several lofty peaks in the ranges near Nadrau that challenge its supremacy. To give a description of the mountain ranges in Viti Levu would be an almost impossible task, they are so numerous, and they divide and subdivide, and connect in a manner that seems to be perfect confusion. The watershed just described, will give an indication of how the land lies in a general way. There is, however, a singular feature connected with this main watershed, which deserves notice. The Siga Toka flows for a long distance in a deep valley, cut down as it were, through the centre of the watershed, and dividing the eastern from the western slope. At Nadrau this valley contracts into a gorge about 800 feet in depth, and so narrow and steep that the sun shines upon the town only between the hours of 11 a.m. and 2 p.m. The town is built at the bottom of the gorge, and on the right bank of the river, which at this place is a roaring torrent about 5 yards wide and 4 feet deep. The rocks on both sides are agglomorate, and it is difficult to say whether the chasm has been cut out by the water, or rent by an earthquake,—most probably the latter.

The scenery on some of the rivers is grand, especially on the Navua, and the upper parts of the Wai ni Mala. In many places these rivers and their affluents have cut out channels for themselves through the rocks and hills, which present many curious and striking features to the Physiographer.

The island of Taviuni lies to the S.E. of Vanua Levu, and is about 21 miles long by an average breadth of 12 miles. It may be said to consist of one mountain ridge, shaped like a pavilion roof, and descending at its ends and sides to the sea. In the centre it rises to an elevation of about 3,000 feet above the sea. On the windward side it is covered with dense forests from the sea to the summit of the mountains. On the leeward side, where settlements are most numerous, a good deal of timber has been felled. There are no large streams on the island. In many instances the water flows underground through caverns, which are numerous, and boils up on the beach, or a short distance out in the sea. There are several extinct craters on it. From the position of the island, its configuration and its dense forests, the rainfall is very great, especially on the windward side.

There are several fine roadsteads on the N.W. side of the island, but no harbours. Vessels do sometimes anchor and ship produce at some places on the windward side, but in general it is an iron-bound coast, unsafe to approach on account of the surf, the high seas, and the wind blowing constantly upon it.

Rabi is a fine fertile island, about 8 miles to the north of Taviuni, and lying between it and Natawa bay in Vanua Levu. It has an area of about 28 square miles. The mountains rise to a height of about 1,500 feet. There is one principal range with many lateral spurs and offsets. Rabi is well wooded and watered, and has many lovely valleys. The anchorage is good, and there are several small but safe harbours. However, the sea surrounding the island is full of reefs and sunken patches of coral, which can only be detected from the mast-head.

Loma-Loma.—The island of Loma-Loma lies to the south-east of Taviuni, in the eastern part of the group. It is about 16 miles long and 3 miles in breadth. A range of hills runs up the centre of the island, from which there are numerous offsets. It is not well wooded, except at the northern end, but is fairly well watered. It has good anchorage inside the reefs, and close to the town of Loma Loma.

Koro.—The island of Koro lies about midway between Taviuni and Ovalau. It rises about 1,500 feet above the sea, and has an area of 58 square miles. It is fairly well wooded and watered, and the scenery in some places is very fine. There is good anchorage for vessels on the northern side of the island, with deep water close to the shore.

Ovalau.—Ovalau lies about 21 miles to the east of Viti Levu, and occupies a central position in the group. It has an area of about 42 square miles. It is very mountainous, rising in some places to an elevation of about 2,200 feet above the sea. There are two principal ranges of mountains, one on its eastern and the other on the western side. From these ranges there are many offsets, which again throw off numerous spurs and crags. The scenery is consequently varied; narrow gorges with valleys opening again into other valleys many times repeated. The central or Lavoni valley, dividing the mountain ranges, runs north and south. The mountains rise in almost perpendicular cliffs on both sides, but their elevation diminishes towards the ends of the valley. The valley opens out on the west side towards Bureta, from which it is entered between two precipices, from three to four hundred feet in height, and about 200 yards apart. The Lavoni valley, seen from an eminence, is one of the loveliest in Fiji. From about 1,000 feet above the sea, to their summits, the mountains are covered with wood. Ovalau is well watered. In the bottom of every valley there is a streamlet, which, in ordinary seasons, flows throughout the year. The scenery is very beautiful in many places, but is not on such an extensive and varied scale as in some of the larger islands.

Levuka, the present capital of Fiji, is situated on the eastern side of the island. It is a straggling town of one principal street, which, extending along the beach almost at high-water level, is bordered on one side by the sea. In this street are the shops, merchants' offices, stores, &c. These occupy nearly all the level ground. The dwelling-houses are perched upon the rocky mountain side above the town, and are approached by steep winding paths, over rocky slopes, or by stair-like ascents. A worse situation could not be chosen for a town. Indeed, except its central position to other parts of the group, it has nothing to recommend it.

CHAPTER X.

Hot Springs—Rocks—Minerals—Soil.

Hot springs are met with in many parts of Fiji. I visited several of them, viz., at Wai Basaga on the Siga Toka, about 5 miles from Fort Carnarvon; one in Rabi; one on the northern shore of Natawa bay (where I believe there are several others); one at Wai Nunu; and the famous springs of Savu-savu. At Vuni-sawan at the head of Natawa bay, there were hot springs which had a wide reputation for curing many kinds of complaints. They were, in consequence, resorted to by large numbers of people, and the usual hospitality to strangers, enforced by custom and tradition all over Fiji, became so heavy that it impoverised those who lived near the springs; therefore the people shut the springs up, after much trouble and labour. The natives pointed out the site of the springs, in the bottom of a muddy creek.

The most extensive hot or boiling springs in Fiji are at Savu-savu, on the south coast of Vanua Levu. They extend for nearly half a mile along the beach, where in many places, a hole scraped in the shingle or black sand, is immediately filled with scalding hot water. The principal boiling springs, three or four in number, are situated about 100 yards from the beach, in the centre of a hollow, which is surrounded by a mound of earth. The water in these boils up to the height of about a foot, with a gurgling sound. After issuing from the springs, the hot water mingles with a stream of cold water which flows within 3 feet of them, and enters the sea by a small creek up which the tide flows. At high water this creek is frequented by the natives for bathing, and the hot water from the springs cooling as it mingles with the water from the sea, baths of any degree of temperature that the body can bear may be had by going a few yards up or down the creek. These baths, the natives say, are efficacious in removing and curing some kinds of complaints. The natives who live in the vicinity cook the most of their food in the boiling water of the springs. The food is put into a basket and covered with straw, and the basket is placed in the hollow or basin where the water boils up. Stones are sometimes laid on the straw to keep the food from being thrown out by the force of the water. The black soil in the vicinity of the springs is incrusted with a white saline substance. There is no vegetation within several yards of them. I believe that no satisfactory analysis, to show the chemical constituents of the water of these springs, has yet been made. The rocks near them are the breccia, so common in Fiji, and a rock resembling shale.

The rocks most common in Fiji are calcareous, like marl and limestone, and volcanic breccia or agglomerate.

These two are about equally abundant, the former uppermost, the latter the underlying rock. Sandstone and shale like rocks also occur at some places, notably in the interior of Viti Levu and Vanua Levu,

Basaltic, trachytic, or porphyritic rocks are not uncommon in some localities, and the lofty mountain peaks (like Voma in Viti Levu), in both the large islands, are composed of them. In the instance of Voma, the basaltic rocks rest upon the agglomerate which was noticed where the ascent of the peak was commenced, at about 1,800 feet below the top. Columnar basalt is rare in Fiji, and the Fijians formerly used its columns, or a portion of them, to mark, or record, certain epochs or events which occurred in the history of a tribe. These columns are generally seen near the sites of the *Devil's Bures* or old heathen temples, and mystical or ceremonial rites may have been attached to them. Near Savu-savu some columnar basalt may be seen in *situ* in the bed of a stream.

In Viti Levu and Vanua Levu, sedimentary or limestone rocks are found on all the mountains. Where absent, denudation, it is conjectured, has been the cause. It may be presumed that the characters of these rocks have in many instances been altered by volcanic heat, so as to resemble rocks of a different origin. A case in point was noticed in the bed of the Wai ni Loa branch of the Wai ni Mala, where the strata had been folded into a vertical position, and exposed to view by the action of the river.

The doubling up of the strata, from whatever cause, and upheavals, may be adduced as reasons for the greater part of the surface of Fiji being hilly and uneven. At Suva, the strata on the sides of the slopes lie at a greater or less angle, while on the top of the slopes the strata assume a somewhat horizontal or unconformable position. This would point to the presumed unconformable strata either having been deposited after the formation of the other strata, or what is more likely, these are the top of the folded strata, *i.e.*, the strata forming the slopes, and from their horizontal position better preserved from denudation. This rock in many places is soft, and crumbles when exposed to the air. It is smooth and very slippery when wet or polished, and is locally known as *Soap-stone*. In other places it is hard, brittle, and shattered to the size of road macadam, often rough from coralline and other sands, pieces of shells, &c., being imbedded in it. These may be noticed on descending the cliffs from the native village of Tamavua to the river, and again at the native town of Kaluba, on the river of the same name, between Suva and the Rewa. At Tamavua and Kaluba, the cliffs are about 300 feet in height. At the last, the river has cut down through this soft rock, and now flows over a bed of hard agglomerate. There are many large caves on the limestone cliffs in the interior of Viti Levu. When coarse-grained, it splits readily into small flags, which the natives, in some localities (interior of Viti Levu), use for paving.

The agglomerate, or breccia, is supposed to be of volcanic origin. It is composed of angular stone, black sand, volcanic ashes, &c. It is very compact, and would make a good building stone, but would be difficult to work. It is a light brown or grayish colour. Where exposed, the edges of the stones in it are rounded off, but they are held most firmly by the binding material, and will break when struck, rather than come out. This binding material is believed to consist mainly of oxide of iron and lime. The rock is found in layers, which vary in thickness and degrees of fineness. But it has no cleavages between the layers. The layers diminish down to a thread-like fineness, and disappear or run into one another. In some of the layers the rock is as fine as some of the rougher building sandstones; in others it is composed of angular stones of various sizes, up to that of a cubic foot. Good samples of these may be seen near Levuka, between Vagadaca and Dribi. In exposed sections, it was noticed that the long stones were lying lengthwise, never perpendicular or across one another. These imbedded stones are of volcanic origin, basalt or trap. To look at the rock *en masse*, where it has been exposed to the weather, and then at the loose disintegrated stones sand, &c., on the beach, or in the bed of a river, one is struck with the similarity of their appearance. This agglomerate is very largely represented throughout the whole of Fiji. The mountains of Ovalau (2,200 feet in height), are composed of it. On the northwestern side of Vanua Levu it forms large cliffs and mountain ridges. In several of these cliffs, seams, or layers of coralline limestone, 6 or 7 feet in thickness, were seen at a few feet above sea-level. In the bed of a stream at Rabi, veins of a trap-like rock were noticed lying between strata of agglomerate; but a doubt exists as to whether these veins are trap, or limestone altered by heat.

At the south-west side of Rabi there is a cliff of sandstone which rises to a height of 20 feet above the sea, and there is a similar cliff on the opposite shore of Kioa island, about 7 miles distant. There is also another on an island at the anchorage or roadstead of Loma-Loma, but possibly it is of coralline origin. The northern part of the island of Loma-Loma is composed of coral limestone, and the southern part, near the town, of agglomerate and basaltic rocks. Some, if not all the islands in this part of the group, are composed of limestone, basaltic, and agglomerate rocks. On the south coast of Vanua Levu, between Savu-savu and Waikava, " Fawn Harbour," the beach and adjacent islands are composed of upheaved coral limestone.

This raised beach varies in width from a few yards to nearly a mile, and rises from sea-level to 100 feet in height. At Mr. Chippendale's estate near Wai Levu, Savu-savu bay, the coral beach has been raised about 30 feet above the sea, and is now covered with soil to a depth of 9 or 10 feet. A gentleman in sinking a well on his property on the Tai Levu coast, about a mile from the sea, found several fine sea shells at a depth of about 11 feet below the surface.

In many parts of Fiji, even on the tops of the high mountains in both the large islands, blocks of coral or coralline limestone may be seen lying about, where nothing remains but agglomerate, or basaltic rocks.

Taviuni is the only island of the group that I visited of a purely volcanic origin, and it would seem also to be a more recent formation than the others. It is entirely composed of scoria, tufa, and basalt, like Mauritius, Samoa, and the Sandwich islands. As already noticed, there are several craters of extinct volcanoes upon the island. This island has been formed above water.

Symptoms of upheavals due to volcanic action, or perhaps to earthquakes, are not wanting in Fiji, and the hot springs are proof that great subterraneous heat still exists below these islands. Unlike Taviuni, the other islands have been formed under water, before being upheaved. That they have been under water before they assumed their present form, is undoubted. In some parts of Viti Levu, and particularly in the centre of Vanua Levu, there are indications, on the surface, of the soil having been exposed to great heat. Still, throughout these islands, there is an absence of what openly appears to have been the craters of active volcanoes, since the upheaval of the land.

Although these islands have been formed by submarine volcanic action, it would now be difficult to point out, with certainty, the positions of the craters. However, it may be assumed that the deep water basins at the anchorages of the islands of Koro, Mango, Rabi, and Savu-savu bay, as well as at the basis of the cliffs on the north-west coast of Vanua Levu, are the sites of craters, and that the perpendicular cliffs at these places formed part of their sides. The hollow of the salt lake, the Lavoni valley in Ovalau, and the site of the town of Namosi in Viti Levu, may be quoted as others.

Graphite, or black lead, is found near na Qarawai in the interior of Viti Levu, and malachite of good quality, according to Seemann, has also been found near Namosi. Copper, but in small quantity, has also been found in Rabi. These metals may yet be found in other parts of Fiji. Flint has been found on the mountains near the native town of na Wasa-kuba, not far from Pickering's Peak. Iron in several of its many forms exists in every part of the group, and numerous fossils will yet be found in its limestone rocks.

Soil.—The soil of Fiji islands is very productive. In Taviuni it consists of disintegrated volcanic rocks, scoria and tufa, mixed with decomposed vegetable matter, one of the richest and most fruitful soils known. Its average depth on this island will exceed 3 feet. In a few places, of small extent, the rocks are covered with a thin coating of soil, but in most places it exceeds 3 feet in depth. Its colour is from dark brown to black. The soil of Ovalau is disintegrated agglomerate and vegetable debris. It is also a fertile soil, and appears to retain moisture better than the soil of Taviuni, which most probably arises from its being more clayey. The decomposed calcareous strata about Suva makes a capital soil when mixed with vegetable matter, and it is surprising to see how trees, reeds, grass, &c., grow on even a few inches of it. To some light porous soils an application of it would act as guano or marl. It retains moisture well. The rock when broken up and exposed to the air, crumbles to powder in a short time; and, in this condition it is capable of producing all kinds of crops. In some localities, of small extent, in both the large islands, particularly in the centre of Vanua Levu, the soil is poor. In these places it is red or white earth or clay, destitute of vegetable matter at a few inches below the surface; but where the surface soil has not been destroyed, a dense growth of trees, bushes, canes, grass, &c. is produced on it. In all the other parts of the group the soil is a mixture of the three first-mentioned kinds, with a large quantity of vegetable matter added. Its texture is loamy, light, and friable. It is not over tenacious of moisture, and water passes readily through it. Unless in the case of the lowest lying lands the sub-soil is well drained. The flats or "bottom land" may be said to be unequalled in fertility. On these lands, crops of sugar cane, cotton, maize, tobacco, &c. have been grown annually for a number of years without manure, and apparently without diminishing the fertility of the soil. What, in Fiji, is generally termed poor land, is only so by the dryness of the locality; and there is not half an acre of land in one place in Fiji so poor as to be unproductive of some kind of useful crop; neither is there any locality in it so dry that grass will not grow for at least nine months of the year on an average of years.

CHAPTER XI.

AGRICULTURAL PRODUCTS—COPRA—COIR FIBRE—
COTTON—SUGAR—TOBACCO—COCOA — COFFEE—
VANILLA—TEA—CINCHONA BARK—RICE.

The present agricultural exports of Fiji are copra, sugar, cotton, maize, tobacco, some arrowroot, and now, perhaps, a little coffee.

Copra, the dried kernel of the cocoa-nut, was exported to the value of £45,908 in 1876; £79,403 in 1877; and £122,194 in 1878. The value of coir fibre produced from the husk of the cocoa-nut was for these years, £2,953, £2,660, and £3,133, respectively. There was also exported during these years cocoa-nut oil to the value of £808 in the first, £941 in the second, and £150 in the third. The value of cocoa-nuts exported in 1876, was £318; in 1877, £461; in £570 for 1878.

The cocoa-nut is extensively grown in the group both by settlers and natives. Young plantations of it are yearly coming into bearing, and new ones are annually being formed. There is yet room for these plantations being extensively enlarged, and that without encroaching on the area which more important products will require. There is now comparatively little cocoa-nut oil made in Fiji. This is principally owing to the high prices which have lately ruled in Europe for copra. An extensive establishment for extracting oil exists at Savu-savu, but the expense of collecting the nuts in small parcels was so great, that the enterprise had to be abandoned. The *Pounac* or refuse of the copra after the oil has been extracted, is a most excellent article for fattening cattle. This is, at present, useless in Fiji, but it is of great value in Europe, and enables the European merchant to give more for the copra than the local oil manufacturer could afford.

But when Fiji is fully settled, and many coffee and sugar cane plantations are in full working order, a local demand for *Pounac* will arise, as in Ceylon and elsewhere, for feeding draught cattle; it will thus become more economical to make oil in Fiji than to sell the copra.

In Ceylon, Seychelles, &c., where the cocoa-nut tree is grown on a large scale, small proprietors have mills, worked by a bullock, for extracting the oil from the copra, which they do to a certain degree of perfection. These mills are sometimes hewn out of stone, but more commonly they are made from the root end of a large tree, of such tough hard wood as *dilo* or *vesi*. The introduction of these simple mills among the natives of Fiji would be of great advantage. They could easily be made in the colony, but one might be required as a sample. This the Government of the colony might undertake.

The fibre yielded by the husk of the nut is, as coir yarn, of great mercantile value for making ropes, matting, &c. The Fijians take so long a time in plaiting their sinnet or coir by the hand, that although done in a superior manner, the amount which the merchants can afford to give for it is disproportionate to the labour expended on it; consequently the natives make it only for their own use. In Ceylon there are simple machines for twisting or spinning coir, worked by the hand. With one of these a man, aided by one or two members of his family, will twine from 50 to 100 lbs. of marketable coir yarn in a day. The introduction of these machines would be of great advantage to the natives of Fiji. They make a great quantity of copra for sale or taxes, but for the want of a simple machine the fibres of the nuts are allowed to rot on the ground. The value of coir per ton equals that of copra, if it does not exceed it, and by the introduction of this simple machine, the Fijian would be enabled to utilise the fibre, and so reap double the benefit he now does from the cocoa-nut.

The number of trees that may be put down on an acre varies from 50 to 100 (from 20×20, to 30×30 feet apart), according to the quality of the soil, and whether the climate of the locality be wet or dry. On an average the trees begin to bear about the seventh year after planting, but a full crop need not be anticipated until the tenth year. This is not owing to absence of flower on the trees, but to a paucity of pistellate flowers, and the inability of these flowers when present to produce fruit while the trees are young. A plantation will continue in prolific bearing for about 80 or 100 years. The average yield per tree per annum is about 100 nuts. About 6,000 nuts yield a ton of copra, and about the same weight of fibre. The value of each is about £14 per ton in Fiji, and these quantities may be produced on about an acre of plantation in a fairly good bearing condition. The estimated cost of clearing the land, planting the trees and attending to them until they begin to bear, is from £15 to £25 per acre, according to circumstances. The after expenses are those of gathering the nuts as they fall from the trees, husking them, and drying the copra. As cattle can be pastured on the grass growing underneath the trees, after they are eight years old, that will pay the keeping of the plantation in good order.

Excepting Viti Levu, the cocoa-nut thrives remarkably well in all parts of Fiji, and the trees are generally as healthy and fruitful as in the best plantations that I have seen either in Seychelles or Ceylon. In Viti Levu the leaves of the trees are attacked by a caterpillar to such an extent as to injure the health of the tree, and prevent its bearing fruit abundantly. The ravages of these catapillars were noticed in some of the other islands, but there they seem to be kept down by birds or insects and prevented from increasing to any injurious extent. The subject ought to be investigated, in order that a remedy may be found and applied; and if this be done the cocoa-nut tree could be increased a thousand fold in the colony. The Government and the Fiji Agricultural Association might take up this matter together.

From what has been said it will be understood that the sugar cane thrives well in Fiji. There are about 24 distinct varieties of the cultivated sugar cane indigenous in the group. These are cultivated by the natives for their own use. They eat the cane in its natural state, and also boil the juice until it attains the consistency of molasses, for sweetening puddings, &c. The wild sugar canes found in the group have already been alluded to.

The manufacture of sugar in the colony has quite recently been commenced, and is yearly growing in importance. The value of the sugar and molasses made and exported during the last three years is as follows: in 1876, £10,433; in 1877, £16,170; in 1878, £18,641. The value of rum made in Fiji and exported during 1878 was £925. Since I left the colony several large mills have been put up and it may be anticipated that in 1880 sugar will be exported to the value of £60,000. These figures show the increasing importance of the sugar industry in Fiji. The total area of land in the group suitable for growing sugar cane is approximately estimated at 1,000 square miles. These cane growing lands are situated in all the principal islands; in the interior of both the large islands as well as near the coast and the flats on the banks of the rivers. In most instances they consist of rich alluvial soil, in the cultivation of which the plough could be used. In the beginning of 1878 a block of about 650 acres of fine cane land, bounded on one side by a navigable river, was sold at auction by Government, for £1 10s. an acre.

When capital has been attracted to the colony, and these cane lands fully occupied and planted with canes, and proper works put up for crushing the cane and making sugar, it may be anticipated that about 200,000 tons of sugar will annually be made in Fiji. The value of this sugar, together with that of molasses and rum, will amount to over FIVE MILLIONS STERLING.

Virgin or plant canes grow and ripen in from 12 to 15 months, and the ratoons can be cut annually. In two instances the growers of the canes have mills on their plantations, and manufacture their own sugar. In all other cases the canes are grown by the planters and sold to the sugar manufacturer at given prices per ton of cane. These prices vary according to the density of the juice :—7s. per ton at 7° Baumè, 8s. 6d. at 8°; 10s. at 9°; 12s. at 10°; 14s. at 11°; and 16s. at 12°. There being few sugar mills the planters are glad to get their canes crushed at any season of the year, often when the density of the juice is at its lowest; even in summer or autumn, the worst seasons of the year for cutting the canes. The ratoons of these canes thus grow during winter and spring, when from the dry cool weather they *will grow least*, and to ripen during the hot wet season, when *they grow most* and ripen least. This not only causes great loss to the planter, but also reduces the value of the exports of the colony. Only when sugar mills become more numerous in the colony will this state of matters be remedied.

The average density of the juice in Fiji is between 9° and 10° Baumé. At the worst season of the year for crushing, the density of the juice is about 5° and at the best between 10° and 11°. These facts, in conjunction with the prices given per ton of cane, at once indicate the loss sustained by the planters. During the summer, autumn, and early part of winter, it took, in 1875-6, 22 tons 4 cwts. of cane to produce a ton of sugar, and from the end of winter, during spring, to the beginning of summer, a ton of sugar was yielded by 15 tons 14 cwts. of cane, or 3 tons of sugar in 47 tons 2 cwts. of cane; while in the former it took 44 tons 8 cwt. of cane to produce 2 tons of sugar, representing money values respectively, of 40*l*. and 60*l*., sugar being valued at 20*l*. per ton on the spot, making thus a loss of 20*l*., or one third.

The expense of cultivation per acre, including cutting, is from 5*l*. for ratoons to 7*l*. for virgin or plant canes. The weight of cane per acre averages 35 tons. From July 1876 to July 1877 the average quantity of cane to the ton of sugar was 16 tons 1 cwt., or about 12 tons when the density was highest, and 20 tons when lowest. There is no disease among the canes in Fiji, nor are they destroyed by insects.

When the disposition of the lands is made, a large portion of cane-growing land will most likely belong to the natives, to grow their *food plants* upon, and produce to pay their taxes in kind, or generally speaking, for the natives to get a living from. If no sugar mills are erected in the neighbourhood of such lands, the natives will be prevented from growing and rendering in kind the crop for which their land is best adapted. In the event of capitalists not accepting a guarantee from the natives, for the erection of sugar mills, Government might give one on their behalf, on approved conditions; the natives of a town or village to cultivate annually a certain number of acres of cane, and, after paying taxes, the surplus of the sugar to be sold, and the proceeds divided among the cultivators.

Coffee thrives remarkably well in Fiji, and its value as an export will ultimately be second only to sugar, and may be expected to amount to between THREE AND FOUR MILLIONS STERLING. The area of land suitable for growing coffee, cacoo, tea, and cinchona is approximately estimated at from 2,500 to 3,000 square miles, and for coffee alone at about 2,000 square miles. Nearly all the islands of the group have land well suited for it, especially Viti Levu, Vanua Levu, Taviuni, Rabi, and Ovalau. The climate is well adapted to the growth of the coffee plant, and so is the soil, limestone being abundant in it. Streams, from which water for pulping could be taken, run in every valley; and in localities where the soil and climate are most favourable, these streamlets most abound.

The coffee plantations I visited were in excellent condition, and the healthy appearance of the plants and their vigorous growth were surprising. At one of nearly 300 acres in extent, the young trees had been topped at a height of $3\frac{1}{2}$ feet, when they were two years old; four months later, when I saw them, the laterals were covering the space between the rows. The plants were put down in rows, 6 feet apart and 6 feet from plant to plant. They were in full flower, and for a first crop the appearance of fruit was all that could be desired, and better than I expected to see. The internodes, between the leaves, were about 2 inches long on an average, which is considered favourable.

At another and older plantation, 5 acres of very irregular but healthy trees, and 5 acres of " 32 months old" trees yielded about 60 cwt. of marketable coffee. The last were planted at a distance of 5 × 5 feet, and under the shade of large trees (the remains of the forest that formerly grew on the land) which had been left standing. These plants were much too close together, and the growth had been so rapid and strong that the lateral branches were overlapping each other about a foot. The internodes, on the branches of these plants, were on an average over 3 inches long, which, as coffee only bears at the axils of the leaves, gives indication of much wood and little fruit. But this is only the result of planting under shade. The 5 years old coffee had also been planted under shade, and some of the trees that had been left standing to protect the coffee had fallen, broken, and destroyed a number of the plants. On the same property, and side by side with these, was another plantation " 20 months old," which had been formed on open ground. The plants averaged about 28 inches in height, and were in full flower for the first time when I saw them. They were in robust health, the leaves of a dark green colour, and gave every indication of yielding an abundant crop. The plants were put down at a distance of 8 × 8 feet, and the side branches or laterals were within a foot of meeting between the rows. The leaves were on an average only an inch and a half apart, a sure indication of abundance of fruit.

A third plantation of nearly 300 acres was being formed, and about 150 acres of it was planted. On a portion of this the seeds had been sown *in situ*, the remainder had been planted with young trees. Both looked promising in the highest degree, the young transplants being about 18 inches in height at the end of 9 months, and robus tand healthy in proportion.

The fourth plantation was, for Fiji, an old one, 9 or 10 years of age. Although it had been abandoned or much neglected, and the plants in many instances allowed to be overgrown by weeds, and broken down by cattle and falling branches, they quickly recovered themselves when care was again bestowed on them, and looked healthy and vigorous, all things considered.

The first of these plantations is on basalt, and the soil is of purely volcanic origin,—scoria and tufa, mixed with vegetable mould. The second is on volcanic breccia, with soil made up of disintegrated particles of that rock, mixed with calcareous or other sedimentary deposits and decomposed vegetable matter. The third is on calcareous and rough sandstone rocks, with soil composed of these rocks disintegrated and mixed with vegetable substances. The fourth plantation was on rich alluvial soil on the bank of a river. At two places on the Wai-dina, and one on the Wai ni Mala, several excellent coffee plants were seen, and their healthy appearance gives great hope of the future success of coffee in the interior of Viti Levu. But a highly commendable action on the part of the Government, in distributing seeds among the natives to form plantations in the interior of that large island, will soon put that matter beyond doubt. A good deal of coffee seed has been imported from Ceylon. Great caution is required in this, lest in importing the seeds, spores of the coffee fungus, which has so disastrously affected the coffee plant in that island of late years should be imported along with them. Indeed, for the welfare of Fiji, an excess of caution in this matter would be laudable. Government might pass an ordinance that all coffee and other seeds, plants of all kinds, soil and packages from Ceylon should be disinfected or purified with sulphur before being landed.* The introduction of this pest would be injurious to Fiji as a coffee growing country. Several (three or four) other plantations are in the course of formation on favourable soil and at suitable situations, and as they are conducted by men of experience in coffee planting, success may be affirmed.

Cotton was extensively grown in Fiji, and it lays claim to producing the best cotton in the world. Fiji is, however, too far from the markets of Europe for her cotton competing in them with that grown in countries more favourably situated. The value of the cotton grown and exported in 1875 was 25,853*l*.; in 1876, 12,022*l*.; and in 1877, 14,140*l*. During this last year 2,000*l*. worth of cotton seeds was exported to Europe. It is still grown by some of the planters, but it will, most likely, be displaced by the sugar cane. From one-half to three-fourths of the cotton now exported

---

* Since these words were written I learn that such a law has been passed by the Legislative Council in Fiji; and since this was printed, I have learned with much regret that the coffee fungus has been introduced with coffee seed from Ceylon. The Government of Fiji is endeavouring to stamp it out by using disinfectants, isolating the infected estates, and destroying the infected coffee plants. There are, comparatively, not many coffee plants in Fiji, and it is to be hoped that the endeavour will be successful. Mauritius, 30th November 1880.

from the group is grown by the Fijians for their taxes. When grown on an extensive scale it still yields a small profit, but the profit is so little that plantations of less than 200 acres scarcely do more than pay labourers' wages. It is otherwise with the Fijians who have the produce for their work. The cotton plantations were mostly small, from 20 acres to over 200 acres in a few instances. These small plantations flourished when cotton sold at about 3s. per lb. in the home market, but when the prices fell, after peace was established in the United States, these small cotton estates had to be abandoned. Were the manufacture of cotton cloth to commence on a liberal scale in either New Zealand or Australia, cotton growing would, most likely, be revived in Fiji. Its cultivation, unlike that of the sugar cane, which requires the investment of a large amount of capital for machinery, and coffee, from which no return is obtained within three years, can be commenced on a small amount of money, and receipts obtained within a year. Therefore, while not so remunerative a crop as either sugar or coffee, its cultivation has great attractions for small capitalists. These hope to make it a stepping-stone which will enable them to enter on the cultivation of another product that will yield a better reward, but which requires a greater amount of capital to begin with than the growing of cotton.

"Sea Island Cotton" is grown by most of the cotton planters. This variety is doubly valuable on account of its seeds, which are worth 8l. per ton on the spot. The variety known as Kidney Cotton is mostly cultivated by the natives. This variety, as a plant, is not only hardier than the *Sea Island* variety, but its cotton is not readily damaged by rain, and is not injured by hanging in the pod several days after reaching maturity.

The climate and soil of Fiji are favourable for growing *tobacco*. However, owing to a general want in these islands of a practical knowledge in preparing the leaves for the market, tobacco is not so extensively cultivated as it could be; but, there is one plantation of it in the group, at Maro, on the western coast of Viti Levu, where the leaf is prepared in a very creditable manner. In some localities, in the interior of Viti Levu, the Fijians grow it to pay Government taxes. The natives also grow it for their own consumption. They cure the leaves by hanging them up to dry, generally in a shaded airy place, which is the only preparation the natives bestow on the leaf. When wanted for smoking they break a small portion of a leaf, dry it on a hot coal, envelope it with a bit of dried banana leaf, and smoke the tobacco in the form of a cigarette, which they call *Seluka*. They have not much to learn in growing the plant, but instructions in the manner of preparing the leaf would benefit them as well as the settlers.

There are many localities in Fiji well adapted for growing cacao (Theobroma cacao), or chocolate plant, and so are the climate and soil. From measures which the Right Honourable the Secretary of State for the Colonies has been pleased to take, the best varieties of this plant cultivated in the West Indies will soon be introduced to Fiji, where they will undoubtedly succeed.

Vanilla will also grow well in Fiji, and its cultivation, from the suitableness of the climate, &c., will certainly be remunerative. But although the plant has been introduced, some time will elapse before a knowledge of the method of fertilising the flowers and preparing the "pods" for the market becomes general in the group.

In the interior of the large islands there are large tracts of land on which TEA and some of the valuable kinds of CINCHONAS (Peruvian bark trees, or quinquina), will most likely thrive.

Several tea plants (in pots) were noticed in Fiji, in the verandahs of settlers' houses. The variety of the tea plant which will best suit Fiji is that known as the *Assam Hybrid*. It might be introduced from India, and, for an inexpensive trial, distributed in small numbers (three or four plants to a town or tribe) among the natives at such places as Nadrau, Babuca, Namosi Vienunga, Namoali, in the interior of Viti Levu. The cinchonas might be dealt with in a similar manner. The varieties of this plant which will do best in Fiji are cinchona succirubra, or *Red bark*, and other sorts which are indigenous in the warmer parts of the Andes. These might be obtained from India in sufficient quantity for a trial, if not from Mauritius and some of the Botanic Gardens in the Australian colonies, where they are grown in collections of economical plants.

Many parts of Fiji are well adapted for the cultivation of RICE. Throughout the country there are numerous old *dalo* or *taro* beds, which only require a little repair to fit them for its cultivation. It would be a most excellent article for alternate cropping with *Swamp dalo*. The one would in a manner prepare the ground for the other. In the wet districts, hill or mountain rice could be cultivated along with land *dalo*. In most localities two crops of swamp rice could be gathered from the same land in a year. The kinds of rice which the labourers now being introduced from India prefer, are the varieties known by the names of *Mooghy* and *Baalam* rice. The PADDY (seed rice) of these varieties might be obtained from India.

CHAPTER XII.

LABOURERS—MARKETS FOR PRODUCE—LAND
TITLES—KINDS OF NATIVE PRODUCE.

Labourers, for working plantations in Fiji, are mostly obtained from the New Hebrides, Solomon Islands, &c. They are engaged for three years. Their wages, exclusive of food, clothes, passage to and fro, is 3*l.* per annum; including everything they will each cost the employer about 12*l.* per annum. At the expiration of the engagement (three years), they are returned to their homes in *the islands* with well filled boxes of *trade* to show themselves and their goods to their friends. They prefer payment in kind to money. They are strong hardy labourers, but utter savages when they land in Fiji. They have to be taught everything, and require a good deal of *breaking in* before they are useful. They are extremely lazy; and the expense of overseeing them is very great; besides, their work is generally badly done. If the supply of these labourers is unlimited, as asserted, they seem to have an unwillingness to go to Fiji, where the demand for them cannot be met. They prefer going to Queensland, where they get double the wages which they receive in Fiji. But money is said to be of no consideration to them, as they do not understand its value. In emigrating they are moved by what to a European would be a whim or notion. There may be some truth in this; but on comparing the quantity of goods which one of them takes from Queensland, with those which another brings from Fiji, there cannot be a doubt as to which of the two countries the majority of these people would prefer to go. Little dependence can be placed upon them preferring either. Their chief reasons for expatriating themselves for a few years are scarcity of food, quarrels, cruel inter-tribal wars, and, above all, a desire to obtain fire-arms.

A number of Fijians are engaged on the plantations. However, as a rule, they do not like to leave home, friends, families, wives, and children, and enter into an engagement to work for a year to the *Papalagai*, on a distant island, and for from 3*l*. to 4*l*. per annum and found. There they only hear of home and friends by occasional rumour. But they have not much objection to work for 1*l*. 10*s*. or 2*l*. per month and food, for a month or two at a time, provided the work is within an easy distance of their dwellings, which they can visit on Sundays or other occasions.

To neither of these could a capitalist depend for labour to work his plantations. It would follow that, when he most wanted labourers, none were to be had. Painfully, such is sometimes the case at present. To put an end to uncertainties on this matter, and to obtain cheap and good labourers, as well as a source to draw them from, that can be dependeded upon to yield a large supply on demand, the Government of Fiji have arranged with the Government of India to get coolies from India. With India to draw upon, no planter, who possesses means, need, in Fiji, fear the want of labourers to cultivate his fields. The engagements with these labourers are that they stay 10 years in the colony, and serve their first employer, at whose request and expense they were introduced, five years.

Their wages does not exceed 1s. per day, all things included. For field work, &c. in the tropics, East Indians are unrivalled. Without them some British colonies, and some foreign ones, would not be in their present flourishing condition. There need be no hesitation in saying that what has been done, and is now being done in these colonies, with coolie labourers, can likewise be done with them in Fiji, which, from its agricultural wealth, crying for development, is bound to become not the least important of Her Majesty's possessions.

Geographically, Fiji enjoys a most enviable position, in regard to a market for her produce, such as no other British (tropical) colony does. Situated within the tropics, she is within 5 days of Auckland, New Zealand; 7 of Sydney; and 9 of Melbourne by steam. The climate of the country is healthy, the soil fertile, rains abundant, and all are favourable in a high degree for the cultivation of tropical products:—indeed, just what are most wanted, sugar, tea, coffee, rice, &c., in Australia and New Zealand. The demand, or market in these colonies, for those products is at present large, and will increase with the population to an unlimited extent. No country is so favourably situated, in all respects, for supplying such products as Fiji.

To those who have capital, and are skilled in the cultivation, &c., of tropical produce, or who can secure the services of trustworthy men as managers with the necessary knowledge, Fiji affords many opportunities for profitable investment.

Again, sugar mills are much needed in many parts of the group by the settlers, who would give security to the sugar maker or owners of the mills for the cultivation of cane to keep the mills in profitable work.

All land titles are given by the Crown, *i.e.*, the Government of the colony. After careful inquiry *Crown grants* are issued as fast as circumstances permit, to the old settlers, who purchased their land from the natives before the sovereignty of Fiji was ceded to Great Britain. Many of these settlers own large tracts of good cane and coffee land, which, from want of funds to cultivate, many of them would gladly sell, at prices from 10s. the acre and upwards according to circumstances of site,—cleared or uncleared land, stocked with cocoa-nut trees or not.

As reference has, in previous paragraphs been made to various kinds of agricultural produce, cultivated by the natives to pay their taxes in kind, it may be of interest to allude briefly to some of them. The principal are copra, cotton, maize, tobacco, and candlenuts. In respect to copra, the plantations or number of cocoa-nut trees owned by the natives is very large. New ones are also being made by them in every part of the colony. The care bestowed on many of these plantations in weeding and general attention leaves much to be desired. Nevertheless the culture of the trees fully equals that of the poorer settlers.

Some of the cotton plantations owned by the natives are well laid down and admirably conducted. The culture of the plants and the cleanness of the ground from weeds are most recommendable. Others, again, are in a semi-neglected state, the ground allowed to be overgrown with weeds, plants growing too thickly (frequently three or more) together, and the knife to reduce the plant to a convenient size for picking the crop, too sparingly employed. The natives sometimes think that, by allowing the plants to grow thickly together and attain a larger size, more produce will be got. Not only so, but that the closeness of the plants will prevent the growth of weeds, and so labour, &c., will be saved.

The plantations of maize are in similar condition to the cotton ones, some of them are well attended to and others indifferently so.

The Fijian does not seem to require teaching in the management of his tobacco plants. He is well aware of the effects of stopping the upward growth of the plants, after they have reached a certain height, on enlarging the lower leaves and perfecting their quality. Experience gained from observation has taught him the best soil and situation for the growth of the plant. His favourite site for a tobacco plantation is that of a house which has been burned. From the plants grown on the few square yards of ground which the house occupied, he will gather upwards of 100 lbs. of leaves.

The facility with which the natives have learned, from the early settlers in the group, the manner of cultivating these articles augurs well for the future, and shows that, were a little care and patience bestowed on teaching them better methods than their own, they would be readily adopted.

The paying of taxes in kind is the best system for the Fijians that could be put into operation, and from the experience which they will gain by it in the cultivation of mercantile products, it may be anticipated that they will become extensive cultivators and producers on their own account.

The *Candle-nut*, *i.e.*, the seeds of the aleurites triloba, is entirely a forest product, to be had for gathering. It grows wild everywhere throughout the group, but it is most abundant on land which has been cultivated or cleared by forest fires. It is particularly common in the sheltered ravines and valleys in the province of Navosa, Viti Levu. During 1875 the value of the candle-nuts exported was £65; in 1876, £1,562 9s.; £3,040 in 1877, and £3,545 for 1878. The value on the spot, in 1877, was a little above £10 per ton, and during this year 300 were gathered and exported. This alone will show the value of the forest products of Fiji, and that the conservancy of the forests in these islands would, besides being generally useful and beneficial to the colony in the way previously indicated, pay its own costs. The *Lauci* or Candle-nut tree, is evergreen. It grows rapidly, and in any situation or soil, produces at an early age, and reaches a height of 60 feet when grown in good soil and sheltered situations. The wood, which is white, soft, and light, is occasionally used by the natives for various purposes, but it is not durable, and therefore valueless for most industrial purposes. The leaves are lobed, or entire, variable in size from 2×3 to 9×12 inches in breadth and length. When young they are densely covered on the upper surface with soft, hoary gray hairs, which gradually fall off, and the leaves become smooth and green coloured with age. The flowers are borne on large panicles which spring from the axils of the leaves or ends of the branches. They are white and not disagreeably fragrant. The fruit is about the size of a small egg. The outside is soft and pulpy and readily decomposes. The covering of the seed or kernel is a hard bony substance, about the twelfth part of an inch in thickness. To get rid of it the natives heat the nuts in a fire and cool them suddenly by throwing water on them; beat them with sticks, or knock them between stones. The *Lauci* yields an excellent blacking or dye, for the manufacture of which an enterprising gentleman has taken out a patent in Fiji. A machine for shelling the nuts would be of great utility, and saving by labour, would cause the nuts to become of greater value, and a larger export of them would ensue. The offer of a premium by Government for a really useful one would stimulate inventors.

In most cases the inhabitants of several towns, and even a tribe, will unite in working a plantation. This answers well when the site of the plantation is in the vicinity of the co-operating townships; but it is different when the villagers have to travel a number of miles to do their share of the work on the plantation. This causes a waste of time, and frequently takes the inhabitants of distant villages from their homes for weeks at a time. Besides, when the plantation is far away from those who cultivate it, it is not so well cared for as it would be if situated in the vicinity of their dwellings. However, the site of a plantation is generally settled by the people themselves, or their representatives in the council of the tribe, the *Bosé vaka Yasana*. It would, perhaps, be better for each village to have its own plantations near the village, and under the Turanga ni Koro, or chief of the village. The Fijian generally works in fits and starts, and either overdoes or does not do enough. Constant daily labour is what he does not like. On the other hand, he is not an habitual idler, and he who does not attend to the affairs of his family and those of his tribe, has not much respect shown to him by his fellow townsmen.

CHAPTER XIII.

STOCK.—PIGS, FOWLS—REMARKS ON THE FAUNA OF FIJI.

Stock, such as sheep, cattle, &c., thrive exceedingly well in Fiji, and every planter has a herd of thriving cattle. These are generally of the best colonial breeds, and frequently expensive cows and bulls are imported from the colonies to improve the stock. To judge from the fine appearance of the animals, the climate of the country favours their growth and increase, and shows that its grasses are sweet and nourishing. In the meantime cattle are almost a drug in the colony and the settlers are looking for a market for them beyond it. When the country is opened up, the oxen will be most useful for draught animals. In general, the land in Fiji is too good for grazing stock upon, or growing maize; that is, crops of sugar cane, coffee, &c., would be far more remunerative, both to the proprietor of the land and to the country. But to the small, not very enterprising capitalist, it is exhilarating to see herds, and, therefore, money annually and surely increasing, at little outlay, and without much trouble. Several of the settlers are rearing herds of Angora goats, and a few of them possess flocks of good sheep.

Pigs, most probably introduced from Tonga, where several were left by Captain Cook, are wild in the forests. In the woods of some localities the domestic fowl runs wild, and is as plentiful as pheasants in a game preserve. Turkeys, ducks, geese, &c., do well in Fiji, and a good turkey may be purchased for a dollar.

The animals indigenous to Fiji are several species of bats or flying foxes and a small rat. A large frog and ten different kinds of snakes are found in these islands. The snakes are harmless and are eaten by the natives. This prevents their increasing in number which, on account of their killing the rats that are becoming numerous and destructive to the crops in some districts, is to be regretted. The natives keep numbers of dogs which they are fond of and kind to. With them the native sportsman hunts wild pigs in the forests, and raises wild fowl, which the hunter knocks down as they fly with stones, or short pieces of sticks, in the throwing of which the natives are experts.

Several kinds of wild ducks are plentiful in the group, and so also are snipe and sandpipers. Wild pigeons are numerous in the forests, and are, like the wild ducks, &c., excellent eating. The golden or orange dove is a most lively bird, whose plumage is gold, or bright orange coloured. Some beautiful parrots or parroquets are also found in the group. The many coloured feathers of these birds and those of domestic fowls, cocks' tails, are used for various ornamental purposes by the natives.

Whales and porpoises abound in the seas which surround the group. These seas also swarm with numerous kinds of fishes, many of which are edible and their quality is most excellent. A few are poisonous, and others are remarkable for their beautiful colours. Sharks also abound in these seas. They are common in deep pools in the rivers; in fresh water, and at long distances from the sea. Fresh water fish, of fine quality and of several sorts, abound in the rivers.

Turtles are found in these seas, and so also is *Beche de mer*, which is annually exported to the value of about £3,500. Shell fish, such as lobsters, &c., are common, and delicious prawns are abundant in all the streams and streamlets in these islands. Some beautiful and rare shells inhabit Fijian waters. Among them may be mentioned the orange cowry, which is only found near Nadroga, on the S.W. side of Viti Levu; oysters are not uncommon, and they attach themselves to the roots of trees, &c., at the edge of the water. Pearl shell is found in small quantities, and in 1877 it was exported to the value of £1,086. Land crabs, some of which are good for eating, are common. One of them, the *Ugavule*, inhabits some of the smaller islands, Qele Levu and Vatuvara, and is said to climb the cocoa-nut trees, remove the husks from the nuts, break the shells, and eat the flesh!

Some kinds of insects, mosquitoes for instance, are common at certain seasons of the year and in some parts of the group; and travellers in Fiji should not forget to take mosquitoe nets with them, as well as mats to sleep upon. Flies are numerous in some localities, and are very annoying.

Some pretty beetles and butterflies are found in the group, and fireflies light up the woods in the dusk of the evenings. Highly curious and pretty leaf and stick insects, different kinds of mantis, are by no means rare. The wings of some of them can scarcely be distinguished from real leaves.

APPENDIX I.

CAOUTCHOUC.

CAOUTCHOUC, or India-rubber, was one of the things which his Excellency Sir A. H. Gordon requested me to inquire into, and determine what trees, shrubs, or climbers yielded it in Fiji :—To ascertain in what manner the Fijians collected the juice, and the nature and habit of the plants which supplied it; to see if they were abundant and where they were growing; and if they could be cultivated for their produce as an article of commerce, and how the caoutchouc could be best collected.

The Fijians name for caoutchouc is *Drega* which means gum or glue that issues from a tree, fruit, &c., when wounded. The term *Drega Kau* is generally applied to trees which have a white milky juice.

The trees, &c., which yield the caoutchouc in Fiji belong to the order apocynaceæ. Several are climbers, and belong to the genera alyxia and lyonsia.

From a variety of causes I cannot with certainty give the names of the species (of climbers) in Fiji; but will send them out with specimens or drawings from England. The trees from which the caoutchouc is obtained are tabernaemontana pacifica, and two if not three species of alstonia, viz., alstonia plumosa a. villosa and another species or variety. Seemann, in his Flora Vitiensis, has doubts about the correct names of them, and as I have no other authority here to refer to, the names will have to be sent from England.

The Tabernaemontana is a medium sized tree, attaining a height of from 30 to 40 feet, with a trunk of about a foot in diameter, and 12 to 20 feet in length.

The leaves are dark green, smooth, thick or leathery, oval, from 3 to 6 or 8 inches long, and from 2 to 4 or 6 inches broad, prominently veined underneath.

The flowers are white, on a two or three times branched panicle at the ends of the branches. The fruit is a berry, about ½-inch in diameter, and 1 inch long, yellow, or reddish yellow when ripe. This tree is found all over the group. It most abounds on alluvial soil, situated in low-lying places. When found in the mountains it is in hollows or flats, where the soil is rich. It is not uncommon, and a good many specimens of it may be seen during a day's journey. It is not gregarious, but three or four trees of it may casually be found near each other.

When the trunk of the tree is wounded, a thin white juice flows freely from it. The Fijians say that the juice curdles, and produces bad caoutchouc.

I could not obtain any caoutchouc from it, though I tried on several occasions. Perhaps this was owing to the small quantity of juice used, and its watery nature. The juice requires a great amount of evaporation before caoutchouc can be obtained from it. On that account it may be dispensed with.

The climbers belonging to the genera Alyxia and Lysonsia are not uncommon in many parts of Fiji. They are most abundant, however, in virgin forests, in the outskirts especially. They attain a large size, and spread to a great distance, overtopping the trees to which they cling, and frequently killing them. Their leaves, on foot-stalks from 1 to 2 inches long, are opposite (verticillate two to four in a whorl in Alyxia) rough or smooth, from 2 to 6 inches broad, and 3 to 9 inches long, ovate or oblong. Flowers white, generally fragrant, rising from axils of the leaves or ends of the branches in large cymes, bi- or tri-chotomously divided.

Fruit of Lysonsia is capsular, that of Alyxia is a soft fibrous drupe. They yield milk abundantly, which readily coagulates. They are rude, large growing plants, requiring supports, and the idea of cultivating them is therefore a questionable one. They are not sufficiently numerous in one part of the forest to render the collection of their juice remunerative.

Specimens of the large-leaved and small-leaved species of Alstonia are enclosed. The former grows to about 30 feet in height, having a trunk which sometimes attains a diameter of a foot or so.

The small-leaved species is in every way a much smaller tree. It seldom exceeds 15 feet in height, and the diameter of its trunk is rarely beyond 6 inches.

This species is most frequently found on the crests of ridges, at elevations of 100 feet above the sea, to the tops of the highest mountains in Fiji. It does not yield juice freely nor abundantly.

Between these two species there is another, or a variety, occupying a middle position in respect to character and habitat. These, however, are so inconstant, that the tree is not always easily distinguished from one or other of its parents, as the two species above-mentioned may be termed.

For the sake of distinction, the large-leaved species may be termed No. 1, the variety No. 2, and the small-leaved species No. 3. While No. 1 is generally found on rich soil at the bottoms of valleys and ravines, and No. 3 on the tops of the ridges, No. 2 occupies the sides of the ravines between them, and makes frequent incursions on the territories of both.

The flowers of the three kinds are cream-coloured, changing to yellow as the flowers grow old.

As specimens of Nos. 1 and 3 are enclosed, they need not be described; and No. 3 need not be further alluded to, on account of the small quantity of juice which it yields.

It is of the large-leaved species and the variety that I will now treat. The follicles, or seed vessels of No. 2, are shorter than those of No. 1: also the tree and its leaves are generally smaller. The seeds of both are from two to three-eighths of an inch long, by one-eighth broad, thickly covered with dark brown hairs, thin and flat, with ciliated edges, which look like fringes when viewed through a lens. The leaves of No. 1 are from 6 to 15 inches long, and from 3 to 6 or 8 inches broad. The leaves of No. 2 are from 3 to 12 inches in length, and from 2 to 6 inches in breadth. Though unequal in size, the trees are equally hardy, and yield juice freely and abundantly. No. 1 would seem to grow more rapidly than No. 2; but this, and the greater size of the leaves, may be caused by the tree growing on rich soil, and in sheltered situations. They may be reckoned one species, whose characters vary according to the position in which the trees may be growing. To this view I am much inclined, because any difference I have as yet seen is in the size of their different parts.

Neither of them is gregarious. They abound in all parts of Fiji, and especially on land which has at one time been cultivated. They are very common at Bua, on the leeward edges of the woods or patches of natural forest which still remain in that district. This may, very probably, be owing to the grass, reeds, &c., having been burnt off the land, and the wind carrying the seed from the adjoining trees on to it. These would germinate after the first rain fell. This suggests a very rapid and inexpensive method of increasing the plant, for useful purposes, in any locality; there being no expense connected with it beyond those of collecting the seeds, burning the grass off the site, sowing the seeds thinly over the ground, and, perhaps, harrowing them in by dragging a few twiggy branches over the surface.

It need scarcely be said that the sowing should take place after the first rains of the season have fallen. In case of greater certainty being desirable, a scratch may be made on the surface of the ground with the foot, a few seeds dropped into it, and the loose soil pushed over the seeds to cover them from the weather. By such simple means, large tracts could be speedily re-wooded with this tree. It only requires the trees to be growing in contiguity to make the collecting of the juice remunerative.

What I saw at Bua confirmed an opinion that I hold regarding the hardiness of the tree. It will grow in the poorest soils, and in the driest parts of Fiji; though, of course, not so well as in a rich soil and moist situation. The tree does not seem to have any particular season for flowering and bearing seeds,— it does so all the year round. This, however, may be a characteristic of the species, not of individual trees. In a locality where the tree abounds, one tree may be found with its flowers in the bud, another in full flower, a third bearing unripe fruit, and a fourth with its follicles bursting, and the ripe seed falling from them. To obtain seed for sowing, the follicles should be gathered when they begin to change colour from green to brown or grey. This is the indication of the seed being ripe. When gathered, the follicles should be spread in a dry, airy room, the floor of which should be covered with paper for the seed to drop upon. The follicles should be frequently disturbed and shaken to take the seed out of them. The seed should be kept in airtight vessels—bottles tightly corked would answer the purpose —until the time for sowing. I am not prepared to say how long the seeds will keep fresh, but judging from analogy, they will not be injured by being kept two or three months. Still, I would say that the sooner they are sown after being gathered the better. The sowing should be done as early in the season as possible, and certainly not later than December. In this a good deal will depend on the rain setting in at an early or late period of the year.

The Fijian name of this tree is *Drega quruquru*. They collect the juice in their mouths, which makes the caoutchouc as adhesive as glue, and of about the consistency and colour of putty. To get the juice, the Fijians break off the leaves from the branches, and collect it as it flows from the petioles and the wounds on the branches caused by the breaking off of the leaves. The branches are next broken off the trees, and each branch is broken up into pieces from 6 inches to a foot long.

As fast as the pieces are broken, first one end of them is placed in the mouth, then the other, till the mouth is full of crude caoutchouc. Several mouthfuls are collected together and squeezed into a round mass or ball. This method of collecting the juice, with the ruthless manner of breaking the trees, somewhat surprised me when I first saw it done. Since then repeated trials in all parts of Fiji have convinced me that the sap or juice does not flow freely by wounding the bark on the trunk of the tree in any way whatever. This is the reason for breaking the branches. The youngest branches of the tree contain most juice. When the old or firm wooded branches are broken very little sap flows from them. When the young branches are broken the sap flows rapidly for a few seconds. It soon coagulates when exposed to the air, and the wound has to be freshened to cause the sap to flow anew. When the branches are broken into pieces of about a foot in length the juice flows from the ends and the pieces are drained almost entirely. A little more may be obtained by breaking the pieces in the middle, but very little. The juice flows from between the bark and the wood, and from the pith, or from between the pith and the wood.

The coagulated juice would seem to have some attraction for the juice in a semi-liquid condition. If a portion of the coagulated juice be applied to the semi-liquid juice adhering to the ends of a broken branch, the slightest touch makes them join firmly. The adhesion is so perfect that the portions will not be separated, and a slight pull takes the semi-coagulated juice clean out of the many fissures or cracks in the ends of the broken branch. To obtain crude caoutchouc from this tree the juice has simply to be collected and worked with the fingers. It requires no other preparation. The juice congeals so rapidly that when collected in dry weather it requires little if any drying. The caoutchouc may be sent to market in balls, or it may be pressed in moulds into long thin pieces, 1 or 2 inches broad and an inch in thickness (more or less) as may be required. Samples of it have been sent to England, and the quality was highly valued:—some of the samples as high as 2s. 6d. per lb., a price equalling that of the best Para caoutchouc.

From the peculiarity of the tree not yielding juice freely by wounding the bark, and the juice being obtained from the young branches, I would suggest that the tree be cultivated and the young branches cut, in the same manner as oak coppice in England. The tree grows rapidly; but I doubt, notwithstanding its rapidity of growth and hardiness, if it would bear having its branches cut annually, or even once in two years. For a beginning I would recommend its being cut every third year. However, cutting annually and biennially might be tried for a number of years as experiments, in order to ascertain what the results in yield of caoutchouc would be; and the effects of cutting each year, and once in two years, on the healthiness and longevity of the trees, as compared with cutting every third year.

The branches may be cut at any season of the year, but the best time for cutting them would be when the sap was most abundant in the branches; and as the tree is an evergreen, and grows all the year, at least in wet localities, the best time will be during the wet season. This is generally the time that the sap is most abundant in the evergreens of the tropics. The sap might be collected at another time, and the result in weight of caoutchouc, from a given number of trees, carefully noted and compared with that obtained from the same number of trees during the wet season.

When the young plants are from 1 to 2 feet in height, their upward growth should be stopped. This may be done by cutting off the top of the leading shoot, or by bending their heads downwards, and keeping them to the ground by placing a stone upon them. Both ways may be tried, but I would commend the latter. By stopping the plants while young no time will be lost in getting them to form coppice stools, and should the plants have grown fairly well, they will likely be fit for cutting to obtain caoutchouc in three or four years after planting. To facilitate the cutting of the branches from which to collect the juice, the stools should not be allowed to grow higher than 5 feet above the ground. The plants may be grown thickly together so as to form dense masses of branches.

When at Rabi, in March last, along with Captain Hill, we made about an ounce and a half of crude caoutchouc from a small tree, the trunk of which did not exceed $1\frac{1}{2}$ inches in diameter. Its head was composed of three or four branches, the average length of which was 2 feet each, total 8 feet of branches from which the juice was obtained. We proceeded in a primitive fashion. The leaves were broken off, and the juice from the petioles was collected on a leaf. The juice from the wounds on the branches, caused by breaking off the leaves, was also saved. The young branches were then broken off the tree, and the juice as it came from the broken ends was allowed to drop on a leaf. These branches were broken again in the middle, but very little juice was obtained by doing so. The juice thus acquired was collected on one leaf, and when worked a little with the fingers it at once formed crude caoutchouc. This caoutchouc was applied to the semi-coagulated juice adhering to the broken ends of the branches. A slight touch was sufficient to join both so perfectly that the whole of the semi-coagulated juice was taken clean away from the ends of the branches and out of all the fissures in the broken stumps on the tree.

Probably the best way to collect the juice will be to cut the young branches into lengths of a foot, and allow the juice to drop from their ends into flat tin vessels. Then break the leaves off the branches and allow the juice as it flows from the leaf stalks, &c., to fall into the same vessel, and apply a portion of the caoutchouc formed to the ends of the broken branches and the stumps on the tree, so as to remove all the juice from them. By such simple means I anticipate that a man to cut off the branches, assisted by two or three boys, will collect 10 lbs. and upwards of crude caoutchouc in a day, provided that the trees be growing closely together.

Working the half congealed juice with the fingers causes it to part with all watery substances, and it is at once caoutchouc, requiring only to be put into shape for the market. Judging from the foregoing example, and it was by no means a fair one, a coppice stool will yield at least two ounces of caoutchouc, and that quantity could be made in about five minutes by a man and two or three boys accustomed to the work. They would soon collect that quantity by their own method of catching the juice in their mouths; and it remains to be proved which of the two ways will be quickest, and give the best result in quantity and quality of caoutchouc.

When once a plantation of the trees has been formed, and the Fijians accustomed to gather and properly prepare the juice, I have every confidence in this matter turning out a success.

As regards a market for the caoutchouc there need be no anxiety. India-rubber is one of those articles that modern civilization cannot do without. The quantity received from America is decreasing, and the supply is not able to meet the demand. India is the only country in which trees yielding caoutchouc are being cultivated and increased.

I have, &c.,
JOHN HORNE,
Director of Gardens and Forests, Mauritius.

APPENDIX II.

SANDALWOOD.

To the Honourable the Colonial Secretary, Fiji.

SIR, Levuka, 31st October 1878.

I HAVE the honour to acquaint you, for the information of his Excellency the Governor, that, when I visited Bua, I made a point of seeing as many sandalwood trees as possible. I therefore requested the Roko Tui Bua to send a man to guide me to where they were growing. This he did, and I was shown 12 trees, and told by the guide that these were all that were known to exist in the district. I was informed in Levuka, that a quantity of sandalwood had been planted by the Roko. I did not see any that had been planted, but if any had been planted, and especially if the plantations had been successful, they would have been shown to me. Besides, it could not have been done without the residents at Bua knowing something about it. The trees seemed to be well cared for; the Roko has a *tabu* on them. Climbers, scrub, &c., are cleared from around them, and they look healthy and promising.

Sandalwood, as everyone knows, was once abundant at Bua. Not only so, but from thence it extended through Macuata to Udu Point. The country from Bua through Macuata to Udu Point, is remarkably well adapted for its growth, but I am thoroughly convinced that no dependence can be placed upon what the natives of these districts say regarding it.

The subject is one which deserves to be strictly and thoroughly gone into. The native tax collector, if nothing better can be done, might be instructed to inquire, and see and examine all the forests in the above named district for himself, in order to ascertain correctly how much there may yet remain of it.

A thorough examination of these is the most essential preliminary step to be taken for the conservancy of the sandalwood. This remark applies also to where the sandalwood is growing in Navosa. To assist in the matter, a drawn up tabular statement is annexed.

The officer who examines the parts where sandalwood may be growing, should fill up the statement, and give answers to the questions in it. The return is filled in for an imaginary sandalwood forest, merely to give an idea of the information required. The returns, when received from different parts, should be compiled into one which should be laid on the council table for the consideration of the Government. When correct information is supplied, Government will be in a position to take proper steps for conserving the sandalwood in the places where it may be growing naturally; or from the few trees that may be growing in one place abandon such a place; or else by planting increase the number of trees so as to make conservancy worth the trouble.

In any scheme of re-wooding for climatic reasons the dry parts of Vitu Levu, and Macuata coast in Vanua Levu, sandalwood, from its value, hardiness, &c., and adaptation to these parts, should hold a foremost place. The tree is a slow-growing one, and compared with some other trees never attains a large size, seldom exceeding 3 feet in circumference at 3 feet above ground. It will probably require about 60 years to reach maturity.

Its wood is dense, hard, and heavy. Its present value in Fiji is about 10l. per ton at the lowest. India is the only place in which it is being preserved and increased. In all the South Sea Islands, if not exterminated, the trees are rapidly decreasing in number. The wood is daily becoming scarcer and dearer, and being a natural product of Fiji, the subject of increasing and making plantations of it, ought to be taken up by the Government. About 600 trees of it, at the least, can be grown and matured on an acre. The trees, when mature, will give an average value of 10s each, at the least; this at present value (which is rapidly increasing) would be 300l. per acre, and, deducting expense of planting, guarding, cutting and transporting the wood to the coast, there would be about 150l. per acre at the least to the good.

In making a plantation of it, the young trees should be put down at the distance apart of 6 × 12 feet = about 600 trees per acre.

The trees that yield caoutchouc could be planted between them at 12 × 6 feet apart, Thus—

× indicate sandalwood, and · caoutchouc trees.

{ × · × · ×
  · × · × ·
  × · × · × ·

This is a small but rapid growing tree, and its produce would pay for the up keeping of the sandalwood trees while the latter were young. As the sandalwood trees increase in size, the caoutchouc trees should be thinned out. The caoutchouc tree grows well, and naturally in the same places as the sandalwood. Of course sandalwood or any other kind of trees could be used instead of the caoutchouc.

In the case of sandalwood being used, one half of the trees would have to be thinned out when they were about one-half or three parts grown, and consequently of no value, on account of the wood not being fragrant. Any other kinds of trees planted would be for the sake of their timber, and as they would also have to be thinned out at an early period of their growth, the timber would be valueless.

The caoutchouc is not a long living tree, and it would be worked out at about the time thinning would be required. The natives could collect the caoutchouc as taxes, or it might be otherwise collected.

In the province of Navosa, which is well adapted for growing sandalwood, the Fijians might plant it instead of paying taxes in produce, or they might be paid for planting it, while at the same time they paid their usual taxes.

In either case the planting would have to be inspected by a trustworthy officer. That being so a good extent per annum would have to be planted in order that the fullest benefit might be obtained for his salary, and the expenses of overseeing reduced to the lowest possible proportion per acre.

The plantation might be extended by annual instalments. The above suggestions may also be applied to Macuata, and the natives in the interior of Vanua Levu could do the planting.

This letter may be read as a continuation of the letter headed na Tua-tua-coka, and dated 19th July last. In addition to the suggestions mentioned in it with regard to planting, &c., it may be mentioned that the weeds, grass, &c., should be burned off the ground before planting by seed sowing, &c., be commenced. Planting should commence after the ground has been watered by the first rains, say in November, and as early completed as possible. When prolonged, the young plants have not time to be sufficiently rooted to withstand the drought of the next dry season, and failure is the result. The seeds preserve their vitality for several months, if kept in a place which is neither too dry nor too damp. To hasten germination the seeds may be steeped in water or mud for a short time before they are planted. After this they must not be allowed to get dry before, or when being put into the ground.

A copy of a regulation suggested for the planting, protecting, and cutting of sandalwood by the Fijians is annexed.

As this report may be thought incomplete without stating the probable results of extensively planting the sandalwood by Fijians or Government, they are here given as follows:—

If 20 men, the inhabitants of one town, Koro, were to plant 20 trees each per annum for 60 years (the shortest estimated time for sandalwood to reach maturity), that would be 20 × 20 = 400 trees planted annually, and 400 × 60 = 24,000 trees at the end of 60 years; and 24,000 ÷ 600 (the number of trees per acre) = 40 acres planted. After the 60th year, 20 trees per man may be felled and sold, and 20 per man planted to replace them. The value of these 20 trees would, at the present value of 10*l.* per ton, be 10*l.* per annum to the heirs or successors of the man who planted them, allowing each tree an average weight of 1 cwt.

In the instance of the Government planting say 100 acres per annum for 60 years, the result would be 100 × 60 = 6,000 acres planted. In 60 years the trees on the 100 acres that were first planted will have attained maturity. They should therefore be felled and brought to market, and 100 acres planted to keep up the stock.

On the 100 acres there will be 60,000 trees at the before given number of 600 trees per acre. These 60,000 trees will weigh on an average 1 cwt. each, and 60,000 ÷ 20 = 3,000 tons, which at 5*l.* per ton will be 15,000*l.* clear profit per annum, after allowing one half the value (5*l.* per ton) for the expenses of planting, preserving, &c.

The amount yielded annually by the caoutchouc would of course be in addition to this. About 7*l.* per ton would be received for the timber standing on the ground, from a purchaser, who would fell and remove the timber at his own risk and expense. This would allow 3*l.* per ton for risk, expense of felling, removing, and profit to the purchaser. Two pounds (2*l.*) of the seven are allowed for the expense incurred by the Government in planting, protecting, &c., leaving 5*l.* per ton net profit.

As the Fijians would perform the necessary labour themselves, they would receive the full value, 10*l.* per ton, for their sandalwood.

It may be desirable for them to plant caoutchouc trees among the sandalwood as above recommended for a Government plantation. If so, they may be directed to put down the trees as in the example given. Otherwise they may be directed to put down the sandalwood trees in rows, 9 feet apart, and at a distance of 8 feet from plant to plant.

They may also be allowed, for the first two or three years, to take a crop of maize, or any other kind of crop that would not injure the young trees, off the land on which the sandalwood was planted.

It may be said "the Government have no right to plant "and grow sandalwood, or any other kind of timber, and sell "it." Such is beside the object, which may be stated broadly as follows:—

The Government have to plant for climatic purposes, and for the preservation of a constant supply of water in the streams throughout the colony. Expenses must therefore be incurred in planting, and in protecting the trees planted from injury or destruction by any cause whatever, and also in reserving, for the same purposes, tracts of natural forests, and preserving the trees on such such tracts from being destroyed or injured in any way.

It is evident that if the trees be not felled when they have reached maturity, and the timber utilized, they will rot on the ground and the timber be lost, and the community will thus fail to derive that benefit from the estate which it ought to receive for the money annually expended in planting and protecting trees on it. By utilizing the mature products the estate would be rendered self-supporting. Thus, while serving the other purposes for which it was upheld, the estate would be yielding to the public the fullest benefit which such kind of estate may be expected to give.

I have, &c.,
J. HORNE,
Director of Gardens and Forests, Mauritius.

The Honourable the Colonial Secretary, Fiji.

SIR, Na-tua-tua-coka, 19th July 1878.

AMONG other things which his Excellency the Honourable Sir A. G. Gordon requested me to report upon before I left Fiji was sandalwood. Accordingly, I visited the sandalwood growing in this district two days ago, and in order that steps may at once be taken for its preservation and extension, I desire that you will bring the matter before His Honour the Acting Governor. Owing to my knowledge of Fijian being limited (and my interpreter and Mr. Langton both unwell), I was unable to get so much information on the spot as I should have liked.*

However, I learned that the sandalwood has been long known to the natives (its value, too, I surmise), and was at one time widely spread over the district.

Many of the trees have been destroyed by fire and Fijian cultivation, and it is now confined to two or three places, which I estimate cover about a square mile each.

Leaving na Suacoka we went up a valley to the left to a Fijian town called na Wasakubu. From this we ascended the hills to the right, and near the top entered the ravine where the sandalwood is growing, through the site of an abandoned township and immense boulders of limestone.

Down this ravine we proceeded for half an hour and arrived at the first sandalwood trees. We found one of them dead; it had been carried down by a landslip. I got it cut up and sent to na Tua-tua-coka, to be transported to Levuka.*

At this place the ravine is narrow and bounded by lands which had once been cultivated, but are now overgrown with reeds. A little further down the ravine gets wider, and branches out to the right and left. In a walk of about half a mile, some 50 or 60 trees, young and old, were counted. These were growing close to the path, and the heads of others were observed among other kinds of trees at a little distance off. Beyond the branch of the ravine on the left there are patches of forest on the sides of the mountain (Koroba), in all of which sandalwood is growing.

The trees among which the sandalwood is principally found are daku; dakua-salu-salu; lewninini; vuga (meterosideros polymorpha); alstonia or the Fijian caoutchouc tree; damanu; acacia richii, *gumu*. The condition of the sandalwood trees is deplorable. Many of them are barked and notched; all of them are broken down by climbers, and choked by useless scrub, &c., and I suspect that numbers have been cut and carried away. The soil on which the trees are growing is decomposing limestone and disintegrating agglomerate.

For their preservation, I would suggest that a trustworthy man be sent to survey the district in which the sandalwood is growing; that the trees, young and old, be counted, and the length and girth of each taken; that the scrub around them be cut and cleared away, and the climbers taken off; that all the dead trees be at once utilized, and the roots also dug up

---

* A large portion of it is now in the Botanical Museum in the Royal Gardens at Kew.

and used; that means be taken to prevent fire; and that seeds be collected and sown where the trees are to grow.

The sandalwood does not transplant readily even when young, and the best way is to sow the seeds *in situ*, two or three together, along with a few chillies to shade the young plants from the sun.

All the district of Navosa is favourable for growing sandalwood, which is one of the most valuable natural productions of the forests of Fiji. The path from the native town of na Wasa-kuba to Tuba-na-sola leads through the part of the forest in which the sandalwood grows.

I have, &c.,
J. HORNE,
Director of Gardens and Forests,
Mauritius.

SUGGESTED (amended) REGULATIONS for planting, protecting, and cutting sandalwood.

As sandalwood is very valuable, it is well that trees of it be planted, preserved, and cultivated; therefore be it enacted:

1. Every man in all towns in the provinces of Bua, Macuata, Navosa, and Ba, directed by the Bose Vaka-Yasana, subject to the Governor's approval, shall in the months of November or December in each year, plant 20 sandalwood trees in a piece of good land selected for the purpose by the Bose vaka-Yasana and the Buli ni Tikini, and such plantations of sandalwood shall be kept clean and in good order and preserved, and such sandalwood shall be the property of the man who planted it, or his heirs, and may be used or sold by him or his heirs when it shall have attained the size mentioned hereafter as the size at which sandalwood may be cut.

2. Sandalwood growing naturally in the forests or planted in a Government plantation is *tabu*, and no person may fell, cut, or remove any sandalwood trees whatever without a license from Government, or if in a private plantation the authority of the owner.

3. It is unlawful to cut any sandalwood tree less than $2\frac{1}{2}$ feet in circumference at 3 feet above the ground.

4. Any person cutting or injuring a sandalwood tree of less dimensions than stated in this regulation, or who has cut sandalwood without a license from the Government or the

authority of the owner of the sandalwood, shall be brought to court and punished by imprisonment for any term not exceeding three months for each tree, besides paying damages, and all sandalwood so cut shall be forfeited to the Crown or to its lawful owner as the case may be.

5. Any person who may either steal or remove sandalwood from any Government plantation or forest without permission of Government, or from the sandalwood plantations of the people without the permission of the owner, shall be punished for theft, in the manner that the laws against stealing direct.

6. Any person who wilfully, carelessly, or thoughtlessly sets fire to any sandalwood plantation or natural forest, or who by raising fire beyond the boundaries of such plantation or forest, and wilfully, carelessly, or thoughtlessly allows such to enter into any sandalwood plantation or forest, shall be deemed guilty of a crime, and shall be brought before a magistrate and punished in such a manner as the law directs.

7. Nothing stated in this regulation shall prevent the offender against any of the foregoing articles or clauses from being prosecuted for high misdemeanour in the Supreme Court should such, upon due enquiry by a magistrate, be considered necessary.

8. The plantations of the natives shall be annually inspected by a Government inspector appointed for the purpose, and the condition of these plantations shall be by him annually reported to Government.

9. Copies of this regulation, translated into Fijian, shall be distributed in all the towns in the districts above mentioned, and after due consideration it shall apply to all those towns or people who possess land suitable for growing and maturing sandalwood.

Regulation No. 6, dated 1878, of the Native Regulation Board is hereby repealed.

JOHN HORNE,
Director of Gardens and Forests,
Levuka, 31st Oct. 1878.   Mauritius.

*TABLE showing the quantity of Sandalwood growing in the District of Navosa.

1. Name of the Place where growing.	1. Estimated Area in Acres.	2. No. of Trees below 6" in circumference.	3. No. of Trees above 6" and below 18" in circumference.	4. No. of Trees above 18" and under 30" in circumference.	5. No. of Trees above 30' in circumference.
Na Wagua *-	50	700	800	600	400

(*continued.*)

6. Names of the principal Trees among which the Sandalwood is growing.	7. Is it overgrown with Climbers, Grass, &c.?	8. Do the Sandalwood Trees appear to be healthy and growing rapidly?	9. What Precautions can be taken to exclude Fire?	10. What would be the Cost to carry out these Precautions per Acre?	11. What watching would be required to prevent theft?
Kau Tabua, Kau Sola, Dakua, Damanu, &c.	Very much.	No, and growing slowly.	Clearing fire breaks through and round sites.	About 3s.	A watchman should be appointed.

(*continued.*)

12. What steps would you propose to be taken for extending the Sandalwood?	13. What is the Nature of the Soil in which the Sandalwood is growing?	14. Is the Site flat or hilly?	15. Have the Fijians planted any Trees, and if so, how many?	16. When were the Trees planted?	17. What is the present size of these Trees?
Planting.	Decomposing sandstone and limestone.	Very steep.	Yes, 15 trees	3 years ago.	10 feet.

(*continued.*)

---

* This *sample* statement is filled in to give an idea of the information required.

18. How far is the site from the nearest Town?	19. What is the Name of the Town?	20. Would the Fijians be willing to plant?	21. If so, how many Plants per annum could a Family put down and attend to.	22. Has there been any Sandalwood cut recently? If so say how much.	23. By whom and for whom was it cut?
10 miles.	Na Tuba. Na Sola.	Yes, if relieved of their taxes.	20	Yes, last year; about 3 tons.	By Fijians of the district under the Buli, and for Government

(continued.)

24. Do the Trees bear Seed freely, and if so, in what Month do they ripen?	25. How many Years does the Sandalwood take to arrive at Mature Stage of Growth?	REMARKS. These may elucidate Answers and refer to Matters not mentioned in the Columns.
Yes, they ripen in December.	About 60 or 70 years, but a great deal depends on the nature of the soil and climate of the locality.	The site, vide Col. 18, is very mountainous, and the Fijians grumble very much at having to cut sandalwood and carry it to the coast. Planting, Col. 20, would seem to take the fancy of the people, especially as they will not have to make taxes during the years in which planting will be carried on. A watchman has been mentioned in Col. 11, if the Government would remit a small amount of the taxes the people of Koro Na Wasa Kuba would gladly undertake it, and be held responsible for the welfare of the plants. Planting has been recommended as the best; the cost would be about 3*l.* for 2,000 plants. As a large area might be planted the annual expense of protection would be nominal, say about 3*s.* per acre per annum. The sandalwood trees are very irregular over the ground, the one half of which they do not occupy. There are also 20 dead sandalwood trees in the forest. These might be utilized. If left much longer where they are the wood will be useless. It will cost 1*l.* per ton to carry it to the coast.

NOTE.—The information required must be for each separate place or forest in which sandalwood occurs. Each place must bear a distinct name or number by which it is or may hereafter be known. The officer who supplies the information must not fill up the columns from memory or hearsay; as correct information is required. He will see and count the trees, &c., himself. If any information be quoted the authority for it should be given.

Signature of officer who wrote this statement.

Signature of officer who counted the trees, &c.

APPENDIX III.

Propositions for a Forest Ordinance for Fiji.

Index.

		Page
Section I. Appropriation of Land for Reserves	-	214
„ II. Mountain Reserves	- -	- 214
„ III. Community Reserves	- -	- 218
„ IV. Timber and Fuel Reserves	-	- 220
„ V. Marsh, River, and Stream Reserves		- 220
„ VI. Survey of Reserves, Plans, &c.	-	- 222
„ VII. Right of Way and Transport	-	- 224
„ VIII. Forest Department; Staff	-	- 224
„ IX. Forest Committee	- -	- 226
„ X. Management of Forest Reserves	-	- 226
„ XI. Sale of Forest Reserve Produce	-	- 228
„ XII. Accounts, Debit and Credit	-	- 230
„ XIII. Branding of Timber	- -	- 230
„ XIV. Offences and Procedure	- -	- 232
„ XV. Power to make Regulations	-	- 234

Propositions regarding a Forest Ordinance for Fiji.

For climatic reasons, and for keeping an abundant supply of water in streams, it is necessary that land be set apart as forest reserves.

Section I.—Appropriation of Land to form Reserves.

1. That all forest reserves be the property of Government.
2. That Government may purchase land to form reserves.
3. That the value of private land to be appropriated for reserves be fixed by arbitration, should principals not agree as to the price.
4. That in the instance of land, the property of any private person being included in area of forest reserve (and owner not disposed to sell it), the owner shall have liberty to fell timber, firewood, &c., but subject to such rules as the conservator of forests (herein-after mentioned) shall lay down, from time to time, as may be required.
5. That felling or injuring trees in such reserves without written sanction of conservator be dealt with as if the land and reserve were Government property.

Section II.—Mountain Reserves.

6. That the sea be the base for all the mountain spurs that terminate on the shore or run parallel to it.
7. That the water in the rivers or streams be the base for all the mountains in the interior of Viti Levu, Vanua Levu, Kadavu, Ovalau (Taviuni?), Koro, and Rabi.

Remarks on Propositions.

Section I.

1. When reserves are the property of private persons, as in Mauritius, it gives rise to a number of troublesome and vexatious prosecutions. The land is looked upon by its owner as his own property, but he is debarred by law from enjoying any benefit from it.

2. When land forming a reserve is Government property, that property may be managed in the manner most beneficial to the public interest.

3. To prevent exorbitant prices being paid for land to form reserves, and give owner what neutral persons consider to be the fair value of it.

4. Permission to fell timber on land forming part of a reserve is what the owner is entitled to, thus enabling him to derive benefit from his land, when for important reasons he might be unwilling to sell it. At the same time, if the land were unwooded, it might cause great injury to the community, were it the watershed of any river or stream supplying a town or thickly populated district with water for domestic use and as a motive-power for machinery.

5. This will prevent such land being unwooded, and will tend to ensure due punishment for reckless destruction of trees.

Section II.—Mountain Reserves.

6. This article does not require explanation, and I would propose that high-water mark, *i.e.*, the place on the shore where the land plants begin to grow, be considered sea-level; a difference of 6 feet might cause an act committed at a certain place to be a contravention or not, depending on whether the base, from which the elevation was taken, were high or low water mark.

7. If the sea were taken for the base of the mountains in the interior of the large islands, it would either cause an unnecessary amount of land to keep as reserves, or it would

Propositions.

8. That the sea-level be the base for the mountains in any other island of the group.

9. That one-third of the elevation of mountains (the highest part) be reserved for climatic purposes, while the lower parts may be cultivated.

10. That if the elevation of any mountain do not exceed 150 feet from the base, there be no reserve on it.

11. That if the elevation of any mountain exceed 150 feet (from the base), the upper part of it, one-third of the elevation be reserved.

12. That any slope of a greater angle than 55° from the horizontal at the base, or more than 35° from the perpendicular at the top, be reckoned part of mountain reserve.

Remarks.

allow land to be unwooded which, for the purpose contemplated, it was desirable to keep in timber. To meet these difficulties, and give a tangible base, the ordinary level of water in streams has been selected.

Where the land plants of a locality cease to grow at edge of the stream, be taken for the ordinary level of the water. Taviuni's place in this article is doubtful. To judge from the conformation of that island, if this rule were applied, Taviuni would be unwooded, and so might be rendered barren, and yet, were one-third of it reserved in timber, too large an area of valuable land would be kept out of cultivation. Of the two evils I am inclined to choose the last, as by the former the whole might be rendered worthless. Its case would be best met by dealing with it separately, but to this there are embarrassing objections.

8. It is scarcely possible to be more than a mile from the sea in the other islands of the group, and none of them contain high mountains or large valleys opening into each other, and the base suggested for them is a natural and fair one.

9. Here, although arbitrary, a line must be drawn, and for the future welfare of Fiji keeping one-third of the elevation of a hill top or ridge in timber cannot be considered too much. This article should be considered conjointly with the next two (10th and 11th).

10 and 11. It is not possible to fix a natural limit, and an arbitrary one has been selected. Considering the numerous ridges on the mountain sides in Fiji, with the fact that unless they exceed a height of 150 ft. from the base no part of them will be reserved, it would follow that nearly all these ridges would be left unwooded. I therefore think the limit should be lowered to 75 or 100 ft., under which there be no reserve. To have all ridges of a less elevation than 75 ft. reserve, and one-third of elevation of those above that height, would cause too much good land to be kept in wood; and to adopt the rule of one-third in all cases would in many instances reduce the part left for cultivation to a mere strip. Much may be said for either way and while I consider a low elevation the best to be adopted, the opinion of others may be worthy of consideration. If in practice the elevation of 75 or 100 ft. be found too low it can easily be raised, but it would cause trouble and expense to lower the limit if one too high were taken.

12. Practically, it is not profitable to cultivate slopes of an angle greater than that mentioned. Including them in reserves will keep them in timber and so prevent landslips.

Propositions.

13. That should the top of a mountain be a plateau 50 ft. from the banks of streams, and 200 ft. from springs and marshes be reserves.

14. That spurs and offsets to mountains and mountain ranges be reckoned part of the mountains to which they belong.

15. That the boundary lines of mountain reserve in all cases follow, *i.e.*, be parallel with the elevation and depression of the mountain ridge or spur.

16. That a base line on one side of a mountain be not the base line for the other side.

17. That in cases of contravention or dispute the elevation be taken from that part of the base line upon which a line passing through the disputed part at right angles to the base line will fall.

18. That timber may be felled on mountain reserves, but previous to such felling the trees be marked by conservator, and that no trees be felled but such as have been thus marked.

19. That mountain reserves be divided and distinctly marked out into "blocks" in which timber may be felled in rotation of a given number of years, to be arranged by conservator and approved by Government.

Section III.—Community Reserves.

20. That there be community reserves for the supply of the native population with wood for fuel, timber for house building, reeds, thatch, &c.

21. That community reserves be subject to rules of felling, &c., suggested for each separate reserve by conservator and approved by Government.

22. That conservator report annually to Government details as to the condition of each and all community reserves.

REMARKS.

13. This rule, while it allows valuable land to be cultivated, will also preserve the streamlets that rise on the plateaux.

14. The idea here is to prevent spurs and offsets of mountains being left unwooded, and only a part of the mountain reserved.

15. The boundary-line of reserve will follow the elevation and depression of ridge. This will prevent unsatisfactory disputes as to where boundary should be, were a line between the tops of two distant peaks taken for the ridge.

16. This is obvious, but were it not stated, injurious decisions might occur.

17. Requires no explanation, but unless mentioned the elevation might be taken from any part of the base line.

18. There are some who maintain that trees ought not to be felled on mountain reserves. This is equivalent to keeping up an estate at great expense, and only deriving half benefit from it. Felling and disposing of mature trees will be utilizing what would otherwise be allowed to waste on the ground.

19. This will facilitate working, insure order, regularity and economy in forest work, and from the produce being found on a limited and compact area, a greater price may be anticipated for it than if it had to be collected from an unknown and unlimited extent of country. The periods (from 6 to 12 years) should be fixed to answer the time when the greatest amount of produce will attain maturity, and so insure the least amount of waste.

SECTION III.—COMMUNITY RESERVES.

20. Were reserves to supply the natives with timber, firewood, &c., not set apart, the wants of the native community would be unprovided for. The result would be the constant plundering of the other reserves when the country was settled and land occupied, and much heart-burning and bad feeling would ensue. A reserve of this kind might belong to one village or it might supply several villages; and from these reserves the natives would have liberty to supply their legitimate wants unquestioned.

21. The principal idea in this is to prevent that abuse of produce in one year, which would cause scarcity or want the next year, or several following years.

22. It is important that Government should know the condition of these reserves, that produce be not abused nor reserve

PROPOSITIONS.

SECTION IV.—TIMBER AND FUEL RESERVES.

23. That there be timber and fuel reserves for the supply of European towns, sugar mills, factories, &c., with wood for fuel, and timber for building and industrial purposes.

24. That in addition to reserves of other kinds named, 33 per cent. of arable land be set apart as reserves for supplying timber for the purposes before stated.

25. That after due consideration, based upon experience, the above per-centage may be lessened.

26. That timber and fuel reserves be also laid out in "blocks" in which timber may be felled in rotation of a given number of years, to be arranged by conservator and approved by Government.

SECTION V.—MARSH, STREAM, AND RIVER RESERVES.

27. That a "marsh" (not salt water) be a place where water stands during nine months of the year.

28. That a "watercourse" be a place in which water flows during nine months of the year.

29. That a streamlet, however small, in which water flows during the year be reckoned a "stream."

30. That marshes, springs, &c. be reckoned part of streams or watercourses.

31. That rivers, streams, marshes, &c., be Government property.

32. That mangrove swamps on the sea-shore, at the mouths or on the banks of rivers be reserves, providing that they exceed one acre in area, or if they exceed 20 yds. in width, by a length of 200 yds. or more.

33. That Government grant access through these reserves to river or sea-shore at certain specified places.

34. That at least 10 ft., measured horizontally, be reserved in trees along each side of streams, and that conservator plant trees on such reserve where they do not already exist.

35. That a board of conservators be appointed for navigable rivers, their jurisdiction to extend from the mouth of the river to the point where navigation ceases.

REMARKS.

damaged for succeeding years. It will be understood that though natives are free to use the produce, they will have no liberty to sell any part of land forming such reserve.

SECTION IV.—TIMBER AND FUEL RESERVES.

23. The utility of these reserves is apparent. They will retain in the colony large sums of money, which, in the absence of such reserves, would annually be sent out of it.

24 and 25. The proportion of land to be set apart for these reserves is large, but in a new country it is difficult if not impossible to forecast with exactness the requirements of a community at a future period, and it would be unwise to commit an error that could not be corrected without much trouble and expense.

26. This is necessary to insure the systematic carrying out of forest operations, and the approval required from Government, will cause the conservator to consider the pros and cons of his suggestions.

SECTION V.—MARSH, STREAM, AND RIVER RESERVES.

Articles 27, 28, 29, and 30 are definitions. With regard to the first and second :—where water remains for nine months of the year without any means being taken to preserve it, it may be concluded that, when means have been applied, the water will remain throughout the year, and the watercourse become a perennial stream.

31. To prevent the damming up of the streams, diverting the water from its course, and to debar any person from using the water so as to injure the community.

32. To prevent these swamps being unwooded and fever thereby engendered. It is assumed that no great harm would ensue from the unwooding of such small areas, and the expense of protecting them, &c. will be avoided.

33. Needs no explanation.

34. The foundation of the subject lies in keeping trees on the land on which the water is collected, as a means to prevent evaporation. Of course this article would be useless to large streams, but to small bodies of water it will be of great service, as they lose more readily by evaporation than large bodies.

35. This article may be deleted.

PROPOSITIONS.

SECTION VI.—SURVEY OF RESERVES.

36. That conservator of forests and Surveyor-General mark boundaries of reserves and set up boundary stones before Crown lands be sold, alienated, or leased.

37. That either conservator or Surveyor-General may depute qualified assistants, who will act together.

38. That the instrument used in taking elevations be a mountain aneroid.

39. That before any reserve be formed, Government be furnished with a detailed report stating reason for reserving, and giving estimated area proposed to be reserved.

40. That along with the report there be drawn out a plan of proposed reserve on a large scale, showing springs, watercourses, streams, &c., and stating of what rivers proposed reserve forms part of watershed.

41. That all reserves be marked out on general plan of island in distinct colours, and that the boundaries of the reserves be carefully and correctly delineated on said plan.

42. That detailed plans of all and each reserve be kept in conservator's office.

43. That such plans be furnished to conservator by the Surveyor-General when the reserves have been surveyed.

44. That each plan of a reserve state area in acres, and give scale.

45. That each plan be signed by the surveyor of the reserve and the draughtsman.

46. That if the plan be a copy, the copy indicate where the original may be seen, and have the names of the surveyor of the reserve, draughtsman, and copyist affixed to it.

47. That without such particulars, no plan of a reserve, or of lands bordering on a reserve, be accepted in a court in bearing evidence.

48. That deputy conservators and rangers be furnished with plans of all reserves within the provinces of which they are in charge.

49. That each reserve be numbered and named, said number to indicate it on map of group, province, and island.

50. That all plans of reserves be duly registered and numbered, the number to correspond with that on the general plan of the province, island, and group for immediate reference and information.

Remarks.

Section VI.—Survey of Reserves.

36. The opposite of this article will show the necessity of marking out the reserves before the disposition of the neighbouring lands. The reason for the two officers working together is to insure that, while the proposed reserve will include all the land that is needful, all land unnecessary for reserve will be excluded.

37. Remarks on this article are needless.

38. The aneroid is mentioned on account of its handiness, and while generally correct, it would require to be tested occasionally. There can be no objection to the theodolite being used.

39. Government being the controlling authority, it is essential that it be in possession of every detail respecting a reserve, in order to form correct opinions on all matters connected with reserves.

40. A plan such as this will not only elucidate report, but it will assist in the formation of correct ideas on the subject, frequently presenting to the mind ideas that otherwise might not occur, and showing the necessity of forming a reserve in the place specified.

41. When a particular reserve is mentioned, this will indicate its situation, and as three kinds of reserves are suggested, a different colour would at once point out each kind on the plan.

42. These plans are necessary, and will be of immense advantage in facilitating forest work of all kinds.

43. Plans being made under superintendence of the Surveyor-General, will ensure their correctness.

44. Necessity of these particulars is evident.

45, 46, and 47. The aim of these articles is to prevent spurious plans being used.

48. This will be of advantage in reporting on any occurrence, and submitting evidence to a magistrate.

49 and 50. The numbering of the reserves and plans, the corresponding of these numbers throughout, and registering the plans, will be found of no small advantage for reference, &c., in all matters connected with forest preservation, and being so, it seems better to provide for this by law, than to leave it to the option of an officer, however intelligent or methodical he may be.

PROPOSITIONS.

SECTION VII.—RIGHT OF WAY AND TRANSPORT.

51. That the right of way through private property to any reserve be secured to Government for inspecting reserves, taking out timber, &c., subject to arrangement regarding expenses of private roads.

52. That the right of floating timber in log and raft down any river, stream, or canal, be secured to Government, subject to arrangement regarding damages to canal gates, wharves, and banks, boats, and all private property whatsoever.

53. That the conservator may, with the sanction of Government, make roads through reserves, construct bridges, &c., and keep the same in repair for the removal of timber, such roads to be indicated on plan of said reserve.

54. That conservator may, with the sanction of Government, remove natural obstructions from rivers for the floating of timber.

55. That conservator furnish details of expenditure on these works, and provide for them in his estimates.

56. That if possible all these works be done by contract.

57. That contractor give security for fulfilment of contract, and proper behaviour of himself and servants, and in case of non-fulfilment of contract, or bad behaviour on the part of contractor or his servants, security be forfeited, and contractor prosecuted, if need be.

SECTION VIII.—FOREST STAFF.

58. That Government appoint an officer to take charge of all reserves, and that officer be styled conservator of forests.

59. That Government appoint as many deputy conservators of forests, rangers, &c., as may be necessary.

60. That the forest staff be permanent and auxiliary, the permanent staff to be as before specified, and the auxiliary staff to be taken on as required, to superintend personally the carrying on of all works in forest reserves, planting operations, &c.

61. That rangers be sworn in before a magistrate.

62. That rangers be acquainted with the Fijian language before being appointed, and that their knowledge of it may be tested by examination.

63. That Government give reward or promotion to deputy-conservators, rangers, &c., for proficiency in the Fijian language, such proficiency to be tested by examination.

REMARKS.

SECTION VII.—RIGHT OF WAY AND TRANSPORT.

51. Were the right of way to a reserve not secured, Government might be debarred from getting access, or having to pay an excessive price for it. The idea is the non-prohibition of Government from using the roads of a private person through land bordering on reserves. Roads to a reserve through adjacent land should be provided in title deeds when the ownership of land is transferred.

52. The reason here is the same as stated in the last, no private person, corporation, or company, should have power to prohibit Government from floating timber down any river or canal.

53. Such provision will prevent misunderstandings, when timber, &c., is offered for sale. Roads through reserves largely enhance the value of the produce.

54. Obstructions, such as rocks, sand-banks, &c.

55. In order that they may be duly considered and provision made for them in the annual budget of the Colony.

56. To keep the officials on forest staff at the lowest possible number. The limiting clause provides for doing the work by the department in the event of its being done improperly, or on unfavourable terms.

57. Requires no comment.

SECTION VIII.—FOREST STAFF.

58. The services of a skilled forester will be essential to organise the department and put it on a proper footing, superintend arrangement of reserves, and keep the whole working in proper order.

59. The reason for this is obvious.

60 and 61. Remarks on these are unneeded.

62. Rangers will have frequent dealings with the natives, and in places where no interpreter is at hand, a knowledge of Fijian will be of importance.

63. To those who have a slight knowledge of Fijian a reward will be an encouragement to increase their knowledge, and it will be to the advantage of the department to give such reward.

Propositions.

64. That conservator, with the sanction of Government, may dismiss any person on the auxiliary staff after due warning.

65. That any officer of the forest department who does not perform his duties in an efficient manner may be dismissed by Government.

66. That Government may authorise conservator to dismiss any official of the department of a lower grade than a forest ranger, but that such official shall have the right to appeal.

67. That Government may summarily dismiss any person employed in the forest department found guilty of having taken a bribe, and that such dismissal do not hinder the Government from proceeding against said person, and also against the party who gave the bribe.

68. That if Government deem it necessary any officer of the department suspected of having taken a bribe, but of which there is not sufficient proof to take before a magistrate, be dismissed from the service, but that said officer be not debarred from proving his innocence by petition to Governor.

Section IX.—Forest Committee.

69. That a forest committee be appointed, of which the conservator be a member.

70. That the governor be judge what matters shall be referred to the committee, and what shall not.

Section X.—Management of Reserves.

71. The conservator send in to Government annually a full and detailed report of all forest operations during the year, report to state, among other particulars, amount received for timber, amount expended on works, amount paid for salaries, how officers have been employed, how contracts have been carried out, how many licenses to fell timber have been granted, amount received for them, rate per tree, rate per cubic foot of timber, number of contraventions, nature of offences against forest laws, and how offenders have been dealt with; in short, every matter with which the department has had to deal during the year.

72. That conservator furnish the Government annually with detailed estimate of income from all sources for the year, under separate and distinct headings, stating from what reserves funds are to be derived, and from what kinds of produce.

Remarks.

64, 65, 66, 67, and 68. Remarks on these are unnecessary.

Section IX.—Forest Committee.

69 and 70. A forest committee (consultative only) will be of the greatest importance, giving weight and authority to decisions on subjects regarding which it was difficult to form opinions.

The conservator will give information to committee on all matters to be considered. Such matters as contracts for forest produce, contracts for works, &c., and utility of entering upon expensive works, might be referred to committee.

Section X.—Management of Reserves.

Articles 71, 72, 73, 74, and 75. Require no explanation.

Propositions.

73. That conservator submit to Government annually detailed estimate of expenditure for the year, under different items, stating where and in what reserve it is purposed to expend money for each kind of reserve and each kind of work separately.

74. That conservator incur no expense for anything whatever, without the previous sanction of Government.

75. That conservator, with the approval of Government, may plant such parts of reserves as it may be thought desirable to re-wood.

76. That before planting be commenced conservator submit to Government scheme of operations, kinds of trees to be planted, number of trees per acre, &c.; cost of every item of expenditure to be incurred under head of planting, watching, preserving, &c., each in detail.

77. That before timber be felled on any reserve conservator submit to Government full, comprehensive, and detailed scheme of operations, proposed expenditure, and expected income, each under separate heads.

78. That trees be planted on the sides of all roads throughout the colony.

79. That trees be planted on the sides of streets in all towns throughout the colony.

80. That conservator be empowered by Government to incur expense in planting these trees and thereafter in attending to them, such expense to be kept separate from forest expenditure.

Section XI.—Sale of Forest Reserve Produce.

81. That timber, &c., on forest reserves be disposed of by auction, estimate, or license to fell per tree.

82. That conservator do not dispose of any forest produce by private bargain without consent of Government

83. That any person entering into contract with the forest department for timber or other forest reserve produce, give such security as may be necessary for the proper carrying out of contract, for correct behaviour of himself and servants, and against all injury to trees, shrubs, or other reserve property by himself, servants, or animals.

84. That in case of non-fulfilment of contract or improper behaviour on the part of contractor or his servants, Government may annul contract, forfeit security, and take possession of tools, implements, &c., in reserves, and belonging to contractor.

85. That such seizure of property do not debar Government

REMARKS.

76. This is an important article, as it requires estimate of expenditure to be laid before Government before planting is commenced. A forester can prepare a specification for planting as easily and correctly as an architect can for a building.

77. Government will thus have full information of works proposed, expected income and expenditure, how forest is to be renewed—whether by natural growth or by planting—and how all work in forests is to be done, &c. This will prevent forest work being carried out in a thoughtless unforeseeing manner, which would lead to the destruction of the forest.

78. Requires no remarks.

79. Unless the management of town roads or streets be in the hands of the municipal authorities.

80. Conservator, being well acquainted with such work, will be the proper person to conduct it, and prepare estimate of expense per mile for doing it. As this expenditure will not be upon forest work, the department should not be charged with it.

SECTION XI.—SALE OF PRODUCE.

81 and 82. Remarks on these unnecessary.

83, 84, and 85. Although these matters might be dealt with by existing laws of Fiji, making provision for them in forest ordinance makes the ordinance more comprehensive and complete.

Propositions.

from proceeding against contractor for default or damage in any court of the colony.

86. That all damages awarded under this head be in proportion to the value of the trees, shrubs, or other reserve property, and the extent of injury done to them.

Section XII.—Accounts, Debit and Credit.

87. That accounts, debit and credit, be kept for each reserve.

88. That proper account of expenditure in connection with each reserve be kept in conservator's office.

89. That all money received for timber and other forest reserve produce, be paid to Receiver-General on approved forms.

90. That all fines, awards for damage to reserve property, and value of articles confiscated, be placed to the credit of the forest department.

91. That proper accounts of all such matters be kept in conservator's office.

92. That the forest department being a source of revenue and expenditure, all accounts connected with it be kept separate from other accounts in the Treasury.

Section XIII.—Branding Timber.

93. That Government have a private mark or marks for its timber.

94. That all wood merchants and holders of licences to fell timber use "brands" or "stamps" for marking all their timber.

95. That all "brands" or "stamps" used for marking timber be registered in conservator's office.

96. That two persons do not use stamps of the same pattern.

97. That it be criminal to counterfeit any timber brand or stamp.

98. That it be criminal for any person to mark timber with another person's stamp.

99. That it be criminal to erase a brand put on timber.

100. That it be criminal for any one to brand the timber of another with his own stamp.

Remarks.

86. For example, a fine of 50*l.* would be a poor compensation for damage done to reserves to the extent of 100*l.* The aim of this article is to guard against this, as experience elsewhere has shown the necessity for it.

Section XII.—Accounts, Debit and Credit.

87 and 88. Keeping accounts in this manner will, in future years, give much satisfaction. It will at once be seen what reserves have not given satisfaction in a monetary point of view, and will cause an inquiry to be made as to the reason, and a remedy to be applied.

89. Needs no remark.

90. This will recompense the forest department for damage done to reserves, instead of allowing fines, &c., for injuries to reserves to be reckoned as ordinary court or police revenue.

91. This will cause conservator to take an interest in seeing that such fines, awards for damages, &c., are paid.

92. In order that when required there be no difficulty in at once ascertaining how the department stands as regards expenditure and income. Accounts of these being kept at Treasury and conservator's office, the one will act as a check on the other, insuring correctness in detail as well as amount.

Section XIII.

93. The idea is that timber, &c., until paid for, is Government property, and a brand on it will indicate that property wherever found. After being paid, a brand of a different kind will be put on the timber.

94. Applies to all timber felled in Government reserves, and will be of great service in preventing timber being stolen.

95. To prevent fraud by altering stamps.

96 and 97. Remarks needless.

98. A might mark the timber belonging to B with C's stamp, either to defraud B or to get C into trouble.

99 and 100. Require no remark.

PROPOSITIONS.

Section XIV.—Offences and Procedure.

101. That any person raising fire in a reserve, and intentionally, thoughtlessly, or carelessly allowing the fire to spread and cause injury to the reserve, be punished by fine or imprisonment according to the extent of the damage, and in addition be made to pay full value for such damage.

102. That any person, master or servant, who may kindle a fire without the boundary of a reserve, and intentionally, thoughtlessly, or carelessly allow the fire to run into and cause injury to the reserve, be punished by fine or imprisonment according to the extent of the injury done, and in addition have to pay full value for such injury.

103. That in the case of fire occurring in a reserve, any person who will not assist in extinguishing the fire when called upon to do so, be brought before the nearest magistrate, and awarded such punishment as the nature of the case may demand.

104. That in the case of fire occurring in a reserve, any person who shall perceive the fire, and not give warning, extinguish, or aid in extinguishing it, be punished by fine or imprisonment as the magistrate shall decide.

105. That illegal felling, injuring, or destroying trees in reserves or on the sides of roads or streets, be punished by fine or imprisonment, according to the nature of the offence and amount of injury done, and offender, in addition to such punishment, have to pay for the injuries done.

106. That any person trespassing in a reserve for an illegal purpose, be punished by fine, and if need be, by imprisonment.

107. That goats, sheep, mules, cattle, horses, pigs, &c. found in a reserve, or browsing on or otherwise destroying trees on sides of roads or streets, be conveyed to the nearest pound, and the owners have to pay for injuries done and expense of pound, or find security for payment, before the animals are given up.

108. That without such impounding of animals, the owners may be brought before the nearest magistrate, who will at once adjudicate, and award such fine or damages as shall meet the exigencies of the case.

109. That in the case of the illegal removing of timber from any reserve, the offender be awarded punishment in proportion to the nature and extent of the offence, and the timber, &c. illegally removed, will, after punishment has been awarded, be restored to its rightful owner, or its value in money should timber be previously disposed of.

110. That all animals, tools, carts, &c. used in the illegal removal of timber be confiscated and thereby become the

REMARKS.

SECTION XIV.—OFFENCES AND PROCEDURE.
Remarks on this section unnecessary.

Propositions.

property of Government, to be disposed of as Government may direct.

111. That any person committing an offence against the forest laws be brought before a magistrate by summons if well known, or warrant if necessary.

112. That if need be, any person committing an offence against the forest laws may at once be arrested and taken before a magistrate.

113. That conservator, deputy conservator, forest ranger, officer of police, or police constable may arrest, or cause to be taken before a magistrate, any person for a breach of the forest laws.

114. That in case of the illegal arrestment of any person, he who made, or caused to be made the illegal arrestment, pay damages, or be imprisoned as the magistrate shall decide.

115. That no case of offence against forest laws be ruled out of court because of informality of proceedings or of information.

116. That any person who shall hinder any officer of the forest department from doing his duty be brought before a magistrate, and awarded such punishment as the magistrate shall deem necessary.

Section XV.—Power to make Regulations.

117. That Government make regulations, with the advice of forest committee, for the carrying out of forest laws.

118. That these regulations, if in accordance with the tenor of forest laws, have the same force as the law.

119. That because any part, article, or clause of these regulations, or the meaning of such be contrary to sense of forest laws, all and every part of regulations be not considered contrary thereto.

120. That all rules and regulations drawn up by conservator, and approved by Government, for the disposal of timber, prices fixed on it, &c., and for the carrying on of all works and contracts for works of all kinds for the benefit of forest reserves, or facilitating the extraction, floating, carting and removing of timber, firewood, or other forest reserve produce from any reserve, have the force of law.

John Horne,
Director of Gardens and Forests, Mauritius.

REMARKS.

SECTION XV.—POWER TO MAKE REGULATIONS.

To provide for the legality of any regulations that may be necessary for carrying out or enforcing the forest laws.

APPENDIX IV.

The Honorable the Colonial Secretary, Fiji.

SIR, Levuka, 8th November, 1878.

I HAVE the honour to enclose for the consideration of Government rules suggested for the felling of timber on lands belonging to Government.

The conditions provide for private parties felling trees on what may be termed a limited scale. However, they can be applied with no alteration to the felling of oversized timber on a forest block of any extent, the right to fell on which was granted, on payment for the trees, to one person. Of course in that case, the permit to fell the trees, as well as the agreement, should state the time in which the trees would be felled and all the timber, &c., removed from the forest. In this instance the number of trees would not be stated in the permit. But the trees should be counted and paid for by tree--the prices varying according to kind, and as per agreement made before the permit was granted. Good security should be given for payment and proper behaviour. The boundaries of the block should be mentioned in the permit. Before any such extensive felling of trees was commenced all conditions, &c., of agreement should be approved by Government. In any case, where timber was being extensively felled, the work should be overseen by a man appointed for the purpose, and who would be always on the spot in the forests, and held responsible for the operations being conducted in a manner beneficial to the Government and the future welfare of the forests. The conditions may to some seem hard and onerous. When carefully examined it will be seen that there is nothing in them but what a conscientious man would of his own accord carry out without them. For such there would not be much need of these regulations, but many consider that trees only grow to be destroyed, and that without benefit to anyone.

The severest regulation will not restrain the actions of such men if the regulation be not firmly and judiciously enforced.

And it is necessary that, before anyone be allowed to put an axe to a tree, they should know what will be the result of misconduct and damage to the forests, and interests of Government.

The suggested regulations are in a crude state, and will required to be put into legal language and form by the Crown solicitor. It is evident that it will not do to allow people to work "*their sweet will*" in a forest; that might cause more injury to the forests, and the Government interest in them, than could be repaired in a hundred years.

I also annex a clause which, until a forest ordinance come into operation, should be added to the suggested conditions under which a permit may be granted. After the forest ordinance becomes a law of the land one of the articles and several clauses provided for the law being duly observed by a permit-holder.

The price per tree has not been mentioned, for, obviously, it will vary according to kind and where the tree may be growing. However, before the tax inspectors were authorised to issue permits, it would be well to give them a general idea of the value of the different kinds, *i.e.*, the sum which the Government expects to receive for each tree of each kind.

I have, &c.
JOHN HORNE,
Director of Gardens and Forests,
Mauritius.

SUGGESTIONS for regulating the felling of trees on Government land.

Permits (licences) may be granted to private individuals, timber-merchants, &c., to fell trees on Government land, subject to the following conditions:—

1. That trees of the following kinds shall not be felled at a less circumference than that undermentioned:—

Tavola,	6 ft. circumference at 6 ft. above ground.		
Vesi,	6 ft.	,,	at 6 ft. ,,
Dilo,	6 ft.	,,	at 6 ft. ,,
Damanu,	4 ft. 6 in.	,,	at 6 ft. ,,
Dakua,	6 ft.	,,	at 6 ft. ,,
Dakua, Salu salu, or Kau Solo,	4 ft. 6 in.	,,	at 6 ft. ,,
Lewininini,	4 ft. 6 in.	,,	at 6 ft. ,,
Kau Tabua,	4 ft. 6 in.	,,	at 6 ft. ,,
Kausia,	2 ft. 6 in.	,,	at 4 ft. ,,
Vaivai,	6 ft.	,,	at 6 ft. ,,
Sagali,	4 ft.	,,	at 6 ft. ,,
Savoo or Cavoo,	4 ft. 6 in.	,,	at 6 ft. ,,
Kau Kuro,	3 ft. 6 in.	,,	at 6 ft. ,,
Mulomulo,	5 ft.	,,	at 4 ft. ,,
Kukulava or Kulava,	4 ft.	,,	at 6 ft. ,,
Tovatova or Vuga, and Vugavuga,	2 ft. 6 in.	,,	at 3 ft. ,,
Moivi,	5 ft.	,,	at 6 ft. ,,
Cibi-cibi,	6 ft.	,,	at 6 ft. ,,
Buabua,	2 ft. 6in.	,,	at 3 ft. ,,
Koka,	5 ft.	,,	at 6 ft. ,,
Nokonoko,	3 ft.	,,	at 6 ft. ,,

Other kinds of trees, and sizes, &c., may be added to this list.

2. That the permits be issued in Ovalau by the commissioner of lands. That in the provinces the permits may be issued by the inspectors of native taxes.

3. That the money be paid to the Treasurer for permits issued in Ovalau. That in the provinces the money be paid to the magistrate or other person appointed by Government to receive money for licences.

SUGGESTED RULES for regulating the felling of trees in Forest Reserves, Fiji, previous to the Forest Law coming in operation.

REMARKS.

1. The different circumferences mentioned in this article are those which the different kinds of trees attain at maturity.

A list giving girth of tree at maturity, under which girth no tree shall be felled, dispenses with marking, and in the absence of a skilled forester to mark the trees that are to be felled, unskilled men can be permitted to fell without risk of injury to forests.

2. The issuing of permits by Commissioner of Lands and Inspectors is meant to be temporary, until a forest department can be organized. These officers, from constantly moving about the country, are better acquainted with the forests than any others, and will thus be able to render efficient service in forest matters.

3. The issuing of, and receiving payment for, permits in the various provinces as well as in Levuka, will be of great advantage to intending out-takers, thus putting them to the least possible inconvenience.

SUGGESTIONS.

4. That the permits be signed by the person who issues them, and by the person who receives the money.

5. That no permit be valid without both these signatures.

6. That neither the Receiver-General in Levuka, nor the magistrates in the provinces issue permits; but no permit will be valid without the signature of one of these officers to certify that the sum mentioned in the permit has been paid.

7. That all payments be made in advance.

8. That each permit state the time for which it is issued, the name of the person in whose favour it is granted, kind of trees he is permitted to cut, the number of them, where they are growing, and the SIZE under which THEY MUST NOT BE CUT.

9. That the permit-holder may not transfer his permit to any person without the sanction of the Government, nor cut trees in any part of the forest, except the one mentioned in the permit, without the authority of Government.

10. That the permit-holder be responsible for all damage caused by fire raised either by himself or his servants, or that may be done by any animals belonging to him.

11. That the permit-holder be responsible for all needless destruction of young trees by his people, either in felling or removing the timber, and for any injury done to any tree by cutting or otherwise, either by himself or his servants.

12. That the permit-holder be responsible for trees being felled either by himself or his servants, within 50 feet horizontal measurement, of a stream, or spring, or of a fresh or salt-water marsh, or within 150 feet perpendicular elevation from the top of a mountain or mountain ridge, unless such trees have been marked for him to fell either by conservator or his duly qualified assistants.

13. That the inspector of native taxes shall be the judge of needless destruction, and from his decisions there be no appeal except to Government in the instance of the award to be paid for damages being thought too high by permit-holder.

Remarks.

4. Accounts of money received for permits at the Treasury ought not to be mixed with accounts of revenue from other sources, but kept a separate item of revenue of themselves.

In this way the amount of revenue and expenditure on forests can easily be got any time.

5. To certify that the permit has been issued and money received by the persons authorised.

6. The Receiver-General and magistrate being unacquainted with the forests might, were they authorised to issue permits, give liberty to fell trees in a place where, for forest purposes, it would be most unsuitable to fell them, and thus unintentionally cause injury that would take many years to repair.

7. To prevent difficulties in collecting the money.

8. Time allowed will vary according to number of trees to be felled, &c., but it is necessary that it should be limited, because, if the work were carried on in a dilatory, trifling manner, it would not pay to look after it.

9. It is desirable that Government should know something about persons who are holders of permits, and if the transfer of permits were uncontrolled, it might give opportunity to dishonest and worthless persons entering and doing damage to forests. Besides, it would open a door for unauthorised proceedings, and tend to subvert all authority in forest affairs.

10. It is essential that the master be held responsible for his servants, else there would be no guarantee against injury done to reserves by the servants of permit-holders.

11. Of course large trees cannot be felled in a thick forest without causing some damage, but permit-holders must know that wanton destruction will not be allowed.

12. This will prevent injury until the forest law be brought into operation. The last clause of this article can then be applied to the authorised felling of such trees as Sagali, useful for piles, or Doga, for firewood, &c., both of which grow in salt marshes, and to the felling of Dakua Salu-Salu, Lewinini, &c., which generally grow on ridges of mountains, and yield timber useful for special purposes. In both places the greatest care is required that too many trees be not felled in one place; for this reason the marking of trees by a responsible Government servant, previous to being felled is proposed.

13. If these matters were to be settled in a court of law, they would be a source of continual trouble, and ultimately the result would be, that the variable interpretations and decisions of the magistrates would be the only recognised

SUGGESTIONS.

14. That no tree (even of the kind named in the permit) be felled unless there be growing within 4 yards of it, more than three young trees of the same kind (or any other kind named in article 1), to replace the one to be felled, and the permit-holder shall be responsible for any tree felled in such situation.

15. That Government will not be responsible to permit-holder, for the kind of tree being in that part of the forest mentioned in the permit, or for the trees being of the required size, but should it be necessary, the permit may be altered by the Commissioners of Lands, or Inspector of Native Taxes for any other part of the forest that may be agreed upon.

16. That the permit-holder remove from the forest all his timber, tools, machinery, animals, and workmen, before the expiration of permit, after which date all timber, tools, machinery, &c., remaining in the forest shall be taken possession of by Government, and for which the permit-holder shall have no claim.

17. That the holder of a permit produce or show his permit at the request of a magistrate, tax collector, inspector, sub-inspector, sergeant or corporal of police, or any European constable, and any permit-holder refusing to show his permit, may be arrested and taken before a magistrate.

18. That forests in which trees are being felled be inspected regularly once a month, once a week, or as often as may be required by an inspector of native taxes, or other person appointed by Government.

19. That the result of such inspection be reported to Government immediately after the inspection has been made.

20. That the permit-holder will, if required, point out to the Inspector of taxes, or any other authorised officer, the place where each tree was felled.

21. The Inspector shall verify, by measuring the stumps, or by measuring the root end of the log (one side of which will, if squared, be taken for one-fourth of the circumference of tree,) that no under-sized trees have been cut.

22. That the permit-holder use a stamp or mark, with which he shall stamp each tree he fells, and the stump from which each tree was cut.

23. That two permit-holders may not use stamps of the same pattern. That it be criminal to counterfeit any such stamp.

24. That a register of these stamps be kept in the Commissioner of Lands' office, and that the stamp used by a

REMARKS.

authority on forest preservation, a result, from experience of the matter, most desirable to avoid.

14. To prevent trees being felled in places where there are no young trees of approved kinds growing, to fill up the place of those to be felled, and thus preserve the forest from being denuded of useful timber trees.

15. Were Government to be responsible for this it might cause trouble, but when it is necessary, another part of forest will be given.

16. To prevent waste of time and expense in looking after cutting or removing of timber, it is desirable that all work be completed within given time, but should circumstances occur unavoidably to prevent permit-holder completing it, he may have an extension of time granted by stating his case to Government, in writing, before the expiration of permit.

17. This is necessary to prevent Inspectors, &c., from being imposed upon by designing persons.

18. It is clear that frequent inspection is necessary, in order to insure the carrying out of rules, otherwise permit-holders would practically be allowed to do as they liked, notwithstanding these rules.

19. That Government may have information of what is being done in forests, and how forest laws are being carried out.

20. To prevent Inspector's time being wasted looking for felled trees.

21. To prevent the felling of under-sized timber.

22. This will prevent timber being stolen, and will aid in the discovery of the party guilty of illegal felling.

23. This article requires no explanation.

24. The registration of stamps is important, as the stamp marked on the permit might be effaced or altered, and then could not be brought as evidence in a court of law.

SUGGESTIONS.

permit-holder be indicated on the back of his permit, and the permit-holder shall not be permitted to use any other stamp without sanction from the Commissioners of Lands.

25. That in all cases of verification the Inspector shall verify, in the presence of the permit-holder, or anyone duly appointed by him, failing his attendance, either himself or by deputy, he shall have no claim against the decisions of the Inspector.

26. That at the expiration of a permit, the person who held the permit shall not be exempt from responsibility as to injuries done to the forest, by himself or his servants during the time the permit was valid, and which were not detected before the expiration of the permit.

27. That a copy of these regulations be given to the out-taker of a permit; another copy with his signature affixed to remain in the possession of the person who issued the permit, and which shall certify that the person taking out the permit agrees to all the conditions under which it is granted, and promises to observe the same faithfully.

28. That if Government deem it necessary, the out-taker of a permit give reasonable security for his general good behaviour and that of his servants, and for the due observance of each article of the regulations by himself and his servants.

29. That a breach of any of these regulations, or of any article of the forest ordinance, will cause the permit to be cancelled, all money and security forfeited, and all timber, animals, tools, machinery, &c., in the forest belonging to permit-holder to be confiscated, and that such seizures and forfeitures do not hinder Government from prosecuting the offender should Government see fit.

I, A———B———, do hereby certify that I will duly observe all articles in the forest ordinance, and the conditions under which I have been granted a permit to cut trees for months in the Government forest at .

Signature of permit holder .

Signature of person who issued the permit

Date .

JOHN HORNE,
Director of Gardens and Forests,
Mauritius.

Remarks.

25. It is necessary that the permit-holder be present to give any information that may be required, and so tend to prevent disputes as to the Inspector's decisions.

26. It would be desirable for the Inspector, accompanied by the permit-holder, to inspect the forest at the expiration of permit, and if the report of that inspection be satisfactory, the permit-holder be free from all further liability.

27. This will prevent the plea of ignorance of conditions or of forgetfulness.

28. I think the necessity of security for good conduct of permit-holder and his servants, and against all damage that may be done will be apparent, and that no permit should be granted without such security.

29. Punishment is not mentioned for infringements of any of the preceding articles separately; this article includes all.

SAMPLE OF PERMIT FORM.

PERMIT to Mr.   to fell and remove 50 Vesi trees from the Government forest at   , in the district of   , during   months from this date.

50 (fifty) Vesi trees at 10s. (ten shillings) per tree = 25l.

No Vesi tree shall be felled whose circumference is less than 6 feet, at 6 feet from the ground. No tree shall be felled in any place or condition forbidden by the forest laws and the regulations under which this Permit has been granted. This Permit is not transferable without the sanction of Government.

Received the sum of
Twenty-five pounds sterling,   Commissioner of Lands, or

Receiver General, or   Inspector of Native Taxes for the District of

Magistrate for the
District of   .   Date of issue,

On back of Permit.

The stamp which the holder of this Permit will use for marking all the timber felled by him is an   stamp, on the head of a   , and the letters or marks are   inch in height.

I hereby certify that the stamp is registered in office of

COUNTERFOIL.

PERMIT to Mr.

to fell 50 Vesi trees in the forest of   , at   , during   months from this date.

Value 10s. per tree = 25l.

Commissioner of Lands.

Inspector of Native Taxes.

On back of Permit counterfoil.

The stamp which the holder of this Permit will use for marking all the timber felled by him is an AB stamp on the head of a hammer or axe, and the letters or marks are one half an inch in height.

247

APPENDIX V.

FIJI METEOROLOGICAL STATEMENTS.

TABLE I.

ABSTRACT OF METEOROLOGICAL OBSERVATIONS,

Taken at Delanasau, Bay of Islands, Bua, Fiji, for the year ending Dec. 31, 1877.

Latitude 16° 38′ S.   Longitude 178° 37′ E.

Height above sea level, 77 feet; distance from sea, 1 mile.

(Extracted from *Fiji Times.*)

1877.	SELF-REGISTERING THERMOMETERS.				RAINFALL.				Average previous six years.
	Mean temp. in shade.	Max. temp.	Min. temp.	Mean daily range.	Ttl. amount in inches.	Greatest daily fall.	No. of days it fell.	Hours of Rain.	
January	80.0	93.8	60.2	13.3	41.70	12.08	23	118	23.46
February	81.3	93.8	72.2	12.0	13.46	2.20	15	54	17.55
March	81.8	94.7	66.8	16.5	7.22	2.74	13	22	25.05
April	80.5	90.8	71.2	12.0	5.65	1.11	14	33	10.49
May	79.7	91.5	64.0	18.3	0.83	0.63	6	4	5.78
June	76.6	89.2	59.5	15.7	0.86	0.46	5	6	2.75
July	74.7	87.4	59.3	15.3	5.52	3.30	9	36	1.38
August	75.0	88.4	56.8	13.5	2.30	0.50	10	24	5.31
September	75.0	89.8	56.3	19.5	0.41	0.37	2	3	4.45
October	78.7	95.8	60.2	20.0	1.29	1.10	4	6	6.93
November	79.9	93.6	68.0	18.0	0.48	0.41	3	4	6.24
December	84.1	97.6	70.8	18.8	0.81	0.81	2	2	9.30
Year 1877	78.9	97.6	56.3	16.1	80.05	12.08	106	312	118.63
1876	79.3	97.0	60.0	16 3	91.36	5.73	135	388	—
1875	79.1	95.5	58.5	15.8	126.64	7.65	146	553	—
1874	79.3	94.1	61.3	15.6	103.48	4.85	165	405	—
1873	78.9	94.5	60.3	15.8	104.10	2.82	151	470	—
1872	78.9	97.5	59.3	15.7	127.03	5.05	180	502	—
1871	79.4	97.7	63.2	15.0	159.51	14.95	180	—	—
Seven years	79.1	97.7	56.3	15.8	113.24	14.95	156	438*	—

* Six years mean.

NOTES on 1877.—

Extreme range of temperature in shade for the year was 41.3, or from 97.6 on 9th December to 56.3 on 14th September.

Highest mean temperature for any 24 hours, 88.1 on 9th December.
Lowest mean temperature for any 24 hours, 69.2 on 9th August.
Greatest daily range, 30.6 on October 9th.
Least daily range, 2.4 on March 31st.
The mean temperature for December, 84.1, exceeds by 2.3 the highest monthly average yet registered at the station, since January 1871.
The temperature in shade in December exceeded 90.0 every day in the month except three.
The rainfall in January exceeded the total rainfall during the remaining 11 months of the year by 2.87 inches.
In 74 hours ending 8 a.m. January 29, 27.32 inches of rain fell causing extensive floods, but little destruction of property; this included 12.08 inches which fell in 24 hours on the 27th.
The only gale during the year occurred on March 31, from S.E., with light rain, thunder and lightening; it was not severe.
There was very little thunder or lightening during the year, particularly during the last nine months. In April thunder was heard on three days, once with lightening; in May on on two days; then none till the 26th November and 30th December.
N.B.—These results have been reduced from observations taken daily throughout the year, at 8 a.m. Thermometers by Casella. Rain gauge by Negretti and Zambra, five inch circular fully exposed.

Further particulars of the drought which prevailed from August 11 to the close of the year:—

1877.		Direction of wind.					Sunshine and clouds.						
	No. of days.	S. to E.	S. to N. (by the east.)	N.	W.	All round.	Days of sunshine throughout.	Three parts sunshine.	Two parts shine.	One part shine.	Overcast.	Rainfall in inches.	Days on which it fell.	Hours of rain.
August 11 to 31	21	18	2	1	0	0	9	2	3	5	2	0.36	3	4
Sept.	30	36	1	0	1	2	11	6	4	6	0	0.41	2	3
Oct.	31	21	5	3	0	2	13	7	7	3	1	1.29	4	6
Nov.	30	21	9	0	0	0	15	8	4	3	0	0.48	3	4
Dec.	31	18	13	0	0	0	14	10	7	0	0	0.81*	1	2
	143	104	30	4	1	4	65	33	25	17	3	3.35	14	19

* This includes 0.63 inches which fell on the 30th.

NOTE.—January 7, 1878.—During the past week only one shower of rain, equal to 0.10 inch fell, making a total for the five months (nearly) 3.45 inches. Yesterday, January 6, temperature in shade rose to 98.6, the highest registered here in seven years.

Wind in general blew pretty strong, but never increasing to a gale nor falling to a calm, except occasionally at night or in the early mornings.

Sun's rays in December intensely hot on some days; on the 24th thermometer for solar radiation, by Negretti and Zambra, with blackened bulb, and placed over short grass, registered 172.0, the extreme limits of the instrument, and on the 26th, 165.7.

On December 24, the salt water in a tidal creek, 5 feet below the surface was 97.0.; near the surface, in blackish water, 95.0

The hygrometer gave some remarkable readings during November and December, showing, for Fiji, extreme dryness of the atmosphere:—On November 18, at 2.30 p.m., dry bulb 93.6, wet bulb 74.8, difference 18.8, deductions dew point 63.5, tension of vapour ·586, relative humidity 37. (Complete saturation of the atmosphere being represented by 100.)

R. L. HOLMES, Observer,
Fellow of the Meteorological Society of England.

(Extracted from *Fiji Times*.)

Month.	Standard Barometer corrected and reduced to 32° Fahrenheit, mean sea level.			Temperature.							Hygrometer.				Rainfall.		
	Mean.	Highest during month.	Lowest during month.	Registering Thermometer in Shade.				Maximum in Sun.	Maximum in Grass.	Mean of Dry Bulb.	Mean of Wet Bulb.	Mean Dew Point.	Mean elastic force of Aqueous vapour.	Total amount in inches.	Greatest fall in one day.	No. of days it fell.
				Mean.	Maximum during month.	Minimum during month.	Mean daily range.									
January	29.824	29.938	29.511	81.3	91.1	69.8	11.0	—	—	81.4	77.0	74.0	.841	12.40	4.43	14
February	29.854	30.088	29.640	82.3	92.2	73.2	12.2	—	—	84.6	76.6	71.3	.769	8.29	1.77	16
March	29.859	29.961	29.595	81.6	90.2	71.7	10.5	—	—	83.9	78.0	74.1	.842	12.67	2.25	21
April	29.852	30.021	29.583	80.6	91.1	72.1	10.6	—	—	82.3	77.1	73.6	.829	4.12	.57	17
May	29.863	30.496	29.826	80.4	91.6	70.0	12.5	—	—	83.1	76.4	71.4	.778	.76	.27	11
June	30.016	30.173	29.838	76.0	89.1	65.4	9.1	147.0	61.0	77.1	71.4	67.4	.672	6.97	2.11	19
July	30.021	30.100	29.865	74.3	85.8	62.4	10.6	146.0	60.7	70.3	70.3	66.1	.614	7.16	1.63	15
August	29.984	30.121	29.890	73.0	83.0	65.0	9.2	142.0	57.5	75.8	68.3	64.3	.646	11.04	2.57	14
September	30.029	30.209	29.874	74.0	82.9	63.0	9.9	134.0	55.9	74.4	68.2	62.9	.575	2.57	2.03	6
October	30.042	30.198	29.884	75.8	85.6	66.3	9.4	151.0	—	75.5	72.2	68.1	.689	2.79	2.03	12
November	30.400	30.138	29.873	78.3	89.3	68.1	10.7	153.6	—	78.0	72.6	66.9	.660	6.70	.01	3
December	29.892	30.058	29.734	81.2	91.2	73.5	11.4	163.8	—	84.3	77.1	72.4	.795	.88	.27	10
Year 1877	29.947	30.299	29.511	78.2	92.2	62.4	10.8	—	—	80.1	73.8	69.5	.722	72.81	4.43	158
„ 1876	29.901	30.189	29.115	79.1	93.0	63.0	12.1	—	—	81.7	75.8	71.9	.782	108.05	6.60	162
„ 1865	—	—	—	—	—	—	—	—	—	—	—	—	—	118.62	5.10	139
„ 1862	29.926	30.407	29.537	80.3	97.6	65.0	—	—	—	—	74.1	—	.781	83.62	5.20	195
„ 1861	29.923	30.636	29.536	78.3	91.9	65.0	—	—	—	—	74.1	—	.795	110.74	4.3	259
Means and extremes.	—	—	—	79.1	97.6	62.4	—	—	—	—	75.4	—	.770	101.93	5.20	134

* Collated from Quarterly Journal of Meteorological Society.

Height of cistern of Barometer above sea level, 19 feet 5 inches.
These results have been reduced from observations taken daily throughout the year at 9 a.m. and 3 p.m. Thermometer by Negretti and Zambra.

Surveyor-General's office, Levuka, 15th February, 1878.

N. LAKE,
Lieut. R.E., Acting Surveyor-General.

TABLE III.

TEMPERATURE OF AIR IN THE SHADE.

Five years monthly means and extremes.
Taken at Delanasau.
(From *Quarterly Journal* of Meteorological Society.)

Month.	Mean.	Max.	Min.	Mean Daily Range.	Mean Monthly Range.
January	80.1	97.7	67.0	13.8	24.5
February	80.0	94.3	68.2	13.1	22.1
March	79.8	97.5	68.5	13.8	24.2
April	79.5	92.4	67.5	14.3	23.3
May	79.1	96.5	60.0	16.4	28.4
June	77.9	92.0	59.3	16.9	28.8
July	77.3	91.7	61.3	17.9	28.1
August	77.4	91.0	58.5	17.3	28.5
September	78.1	92.8	61.7	15.9	26.3
October	79.4	94.5	62.0	16.1	27.5
November	80.0	94.6	64.0	16.6	27.6
December	80.7	94.4	68.5	14.7	23.6
5 years means and extremes.	79.1	97.7	58.5	15.6	26.1

Greatest daily range, 29.5°, May 19, 1872.
Least daily range, 3.2°, April 3, 1871.
Highest mean temperature in 24 hours 80.7°, December 16, 1874.
Lowest mean temperature in 24 hours 65.9°, August 3 1873.

TABLE IV.

Rainfall at Qara Walu, Taviuni, South.—1877.

(Extracted from *Fiji Times*.)

	Total amount in inches.	Amount that fell in the day.	Amount that fell in the night.	Number of days rain fell.	Maximum in one day.	Date.	
January	27.23	12.58	14.65	24	7.73	On the	28th
February	17.93	6.82	11.11	21	3·34	,,	21st
March	54.30	22.02	32.28	21	11.82	,,	31st
April	39.86	21.65	17.21	22	9.40	,,	1st
May	5.38	0.52	4.86	11	3.15	,,	18th
June	23.90	10.67	13.23	18	5.18	,,	5th
July	18.58	6.09	12.49	26	7.04	,,	10th
August	34.25	14.49	19.76	22	5.42	,,	5th
September	1.95	0.74	1.21	13	0.50	,,	1st
October	18.24	9.59	8.65	18	5.48	,,	28th
November	2.69	2.01	0.68	17	0.46	,,	7th
December	8.26	2.14	6.12	15	3.18	,,	18th
Total	251.57	—	—	228	11.82	—	
1875	212.37	—	—	230	7.16	—	

The above rainfall at Qara Walu, Taviuni, South, for the year ending December 31st, 1877, was taken daily for the previous 24 hours. $2\frac{1}{2}$ miles from the sea; height above sea level, 564 feet by Royal Engineer's survey.

J. Newall.

TABLE V.

STATEMENT OF WET, SHOWERY, AND FINE DAYS ON THE REWA RIVER.—1877.

(Extracted from *Fjii Times*.)

	Wet.	Showery.	Fine.
January	6	10	15
February	2	10	16
March	8	6	17
April	3	14	13
May	0	4	27
June	13	9	8
July	3	8	20
August	7	5	19
September	0	1	29
October	7	4	20
November	0	3	27
December	3	3	25
	52	77	236
1877	52	77	236
1876	64	79	223
1875	56	77	232
1874	71	83	221
1873	42	104	219
1872	65	85	216

ED. GRAHAM.

Wai ni Sasa, Rewa River, 25th January 1878.

TABLE VI.

RAINFALL.

Results of Observations for five years ending 31st December 1875, at Delanasau.

(Extracted from the *Quarterly Journal* of the Meteorological Society.)

Month.	Mean Monthly Fall.	Greatest Fall in 24 Hours.	Mean Number of Days on which Rain fell.	Hours of Rain Mean for Four Years only.
	In.	In.		
January - -	25.69	10.52	23	96
February - -	16.76	4.20	20	69
March - - -	25.74	14.95	22	91
April - - -	11.25	4.74	18	41
May - - -	5.78	2.10	11	28
June - - -	3.17	3.90	7	12
July - - -	1.64	1.53	5	8
August - - -	5.69	2.67	10	23
September - -	4.86	3.30	12	19
October - - -	7.16	2.80	13	30
November - -	7.08	5.52	11	25
December - -	9.33	2.90	18	40
Year - -	124.15*	14.95†	170‡	482§

* Mean yearly fall for five years.
† Greatest fall in 24 hours, March 19, 1871. Greatest fall in one month (March 1875) 38.84 inches. Least fall in one month (July 1872) 0.36 inches.
‡ Average number of days on which rain fell; mean for five years.
§ Average number of hours of rainfall for four years, equal to 20 days two hours.

TABLE VII.

HYGROMETRICAL RESULTS DEDUCED FROM DRY AND WET
BULB THERMOMETERS, READ DAILY AT 1 P.M. THROUGHOUT
THE YEAR 1875.

(Extracted from the *Quarterly Journal* of the
Meteorological Society.)

Month.	Mean Temperature of			Pressure of Vapour	Relative Humidity (Saturation = 100.)
	Air.	Evaporation.	Dew-Point.		
	°	°	°	In.	
January - -	86.5	79.5	75.0	.868	69
February - -	85.4	78.9	74.7	.859	70
March - -	82.6	78.5	75.8	.891	80
April - -	84.9	78.1	73.7	.823	69
May - -	86.1	78.3	73.2	.817	65
June - -	84.9	76.9	71.7	.777	65
July - -	84.7	74.6	68.0	.685	57
August - -	84.7	74.4	67.7	.678	56
September -	84.3	75.9	70.4	.743	63
October - -	86.2	74.0	66.1	.642	51
November -	87.1	76.9	70.4	.743	58
December -	87.7	77.8	71.5	.772	59
Mean -	85.4	77.0	71.5	.775	63

The lowest relative humidity noticed was at noon, on 26th
November 1875, namely :—

Air.	Evaporation.	Dew Point.	Vapour Tension.	Humidity.
°	°	°	In.	
89·2	72·0	61·2	·541	39

This occurred at a very dry period, the dry season having
been very much extended. During the three weeks previous
only 0·24 inch of rain had fallen, and strong, at times boisterous
southerly winds had prevailed almost without cessation, and
with, in general, a clear sky. This low amount of relative
humidity may be described as very exceptional in Fiji.

APPENDIX VI.

List of Plants found in Fiji.

Alphabetical List of Plants (Ferns and their allies excepted), which, up to date, have been found in Fiji. The species marked by an asterisk are endemic, or peculiar to Fiji islands. Those species that, at present, are regarded as new are indicated by sp. n. and N.S. It is likely that several of these may be merely varieties of old and well known species. The figures in brackets ( ) are the numbers attached to the specimens, to identify them, in the Royal Herbarium at Kew. Var. means variety. The interrogation note (?) before a species signifies that such species has, most likely, been introduced.

Abrus precatorius.
Abutilon sp. n. (378).*
Acacia sp. n.*
„ laurina.
„ Richii.*
Acalypha sp. n. (1,001).*
„ boehmerioides.
„ consimilis.*
„ denudata.*
„ anisodonta.
„ „ subsericea.
„ „ subvillosa.
„ grandis.
„ insularia.
„ „ flavicans.
„ „ glabrescens
„ „ stipularis.
„ „ villosa.
„ latifolia.*
„ repanda.*
„ „ var., n. (35).*
„ rivularis.*
„ Wilkesiana.
„ lævifolia.*

Achyranthes aspera.
Acicalyptus (Calyptranthes) sp. n. (251).*
„ eugenioides.*
„ myrtoides.*
„ Seemannii.*
Acronychia sp. n. (1,114).*
„ petiolaris.*
Adenosma triflora.
Adenostemma viscosum.
Afzelia bijuga.
Agalma vitiensis.*
Agation (Agatea) sp. n. (67) *
„ „ violare.*
Ageratum conyzoides.
Aglaia basiphylla.*
„ edulis.
„ multijuga.*
„ sp. n. (858).*
„ sp. n. (344).*
„ sp. n. (1,040).*
Aleurites triloba.

Alpinia sp. n. (593).*
„ sp. n. (879).*
„ sp. n.*
„ Boia.*
„ sp. n.*
„ vitiensis.*
Alphitonia excelasa.
„ „ var. n. (655).*
Alocasia indica.
Alsodeia Storckii.*
Alsomitra sp. n. (1,043.)*
(?) Alstonia sp. n. (857).*
„ sp. n. (861) *
„ sp. n.*
„ plumosa.
„ villosa.
„ viliensis.*
Alyxia sp. n. (343).*
„ bracteolosa.
„ „ microcarpa.
„ „ angustifolia.
„ „ parvifolia.
„ scandens.
„ stellata.
Amaranthus melancholicus.
„ „ tricolor.
„ paniculatus.
„ „ cruentus.
„ verides.
Amaroria soulameoides.
Amomum sp. n. (248).*
„ cevuga.
Amorphophallus campanulatus.
Anacardium sp. n. (1,050).*
Andropogon aciculatus.
„ refractus.
Anæctochilus longiflorus.*
Aneilema vitiensis.
Aniseia multiflora.
„ uniflora.
Amplectrum ovalifolium.*
Antiaris Bennettii.

Antidesma sp. n. (491).*
„ sp. n. (692).*
„ pacifica.*
Appendicula sp., N.S.*
„ bracteosa.*
Araliaceae sp. n.*
Ardisia sp. n.*
„ capitata.*
„ grandis.*
„ Storckii.*
„ vitiensis.*
„ sp. n. (180).
„ sp. n. (429).*
„ sp. n. (639).*
„ sp. n. (883).
„ sp. n. (946).*
Aristilochia sp. n.*
Arthropanax fruticosum.
Artocarpus incisa.
Ascarina lanceolata.
Astelia sp. n. (1,028).*
„ montana.
Astronia fraterna.
„ Pickeringi.
„ „ vitiensis.*
„ confertiflora.*
„ robusta.*
„ tomentosa.*
„ sp. n. (616).*
„ sp. n. (1,038).*
„ sp. n. (1,039).*
Astronidium parviflorum.*
„ Storckii.*
Baccaurea Wilkesiana.*
„ Seemannii.*
„ stylaris.*
„ sp. n. (447).*
„ sp. n. (515).*
„ sp. n. (981.)*
Badusa corymbifera.
Bakeria vitiensis.*
Balanophora fungosa.
Barringtonia speciosa.
„ edulis.*

Barringtonia racemosa.
„   sp. n. (619).*
(?) Batatas edulis.
(?)   „ paniculata.
(?) Bauhinia tomentosa.
Begonia sp. n. (618).*
Bidens pilosa.
Bischofha javanica.
(?) Bixa orellana.
Blumea Milnei.*
„   virens.
Boehmeria platyphilla.
„   „   (virgata).
„   sp. n. (57).
„   Harveyi.
Boerhaavia diffusa.
Brackenridgea nitida.*
„   sp. n. (513).*
„   sp. n. (570).*
„   sp. n. (897).*
Broussonetia papyrifera.
Brucea quercifolia.*
Bruguiera Rheedi.
Buchanania florida.
Bulbophyllum longiflorum.
„   sp. n. (730).*
Cæsalpinia Bonducella.
„   Bonduc.
Calamus s.p.
Calanthe hololeuca.*
„   ventilabrum.*
„   sp. n. (285).*
„   sp. n. (876).*
Calophyllum inophyllum.
„   Burmanni.
„   spectable.
„   sp. n. (43).*
Calonyction speciosum.
„   comosperma.
Calyaccion tinctorum.
Calycosia Hunteri, N.S.*
„   Milnei.
„   petiolata.*
„   publiflora.*

Calycosia sp. n. (40).*
„   sp. n. (279).*
„   sp. n. (501).*
„   sp. n. (534).*
„   sp. n. (593).*
„   sp. n. (644).*
„   sp. n. (705).*
„   sp. n. (775).*
„   sp. n. (789).*
„   sp. n. (644).*
Canavalia obtusifolia.
„   sericea.
„   turgida.
Canaga odorata.
Canarium sp. n.*
„   vitiense.*
„   sp. n. (686).*
Canthiopsis odorata.*
Canthium sessilifolium.*
„   barbatum.
„   flavidum.*
„   odoratum.*
Canna indica.'
Capparis Richii.*
(?) Capsicum frutescens.
Carapa moluccensis.
„   obovata.
Cardamine sarmentosa.
Cardiospermum halicacabum.
„   microcarpum.
Carruthersia scandens.*
„   sp. n. (967).*
„   sp. n. (460).*
Carumbium nutans.
Cascaria accuminatissima.*
„   disticha.*
„   Richii.*
„   sp. n. (149).*
„   sp. n. (365).*
„   sp. n. (454).*
„   sp. n. (456).*
„   sp. n. (640).*
(?) Cassia lævigata.
„   glauca.

(?) Cassia, occidentalis.
(?)   ,,   obtusifolia.
    ,,   sophora.
    ,,   sp. n. (648).*
Cassytha filiformis.
Casuarina equisetifolia.
  ,,   nodiflora.
Caturus pelagicus.*
Celastrus Richii.*
Celtis Harperi, N.S.*
Cenchrus anomoplexis.
Centotheca lappacea.
Ceratophyllum demersum.
Cerbera lactaria.
Chætacanthus repandus.
Chailletia vitiensis.*
Chenopodium ambrosioides.
Chloris punctata.
Chrysoglossum vesicatum.*
Cinnamomum pedatinervium.*
  ,,   sp. n. (85).*
  ,,   sp. n. (99).*
  ,,   sp. n. (832).*
  ,,   sp. n. (862.)*
  ,,   sp. n. (872).*
  ,,   sp. n. (974).*
(?) Citrullus vulgaris.
(?) Citrus Limonum.
(?) ,, Aurantium.
(?) ,, vulgaris.
(?) ,, Decumana.
Claoxylon echinosperma.*
  ,,   fallax.*
Cleidion Vieillardi.
  ,,   vitiensis.
Clematis Pickeringi.
Clerodendron Arthurgordoni, N.S.*
  ,,   inerme.
  ,,   Le Huntei, N.S.*
  ,,   sp. n.*
Cocos mucifera.
Codiœum variegatum.
  ,,   ,,   var. 313.

Codiœum variegatum var.(1)
  ,,   ,,   ,, (2).
  ,,   ,,   ,, (3).
  ,,   pictum.
  ,,   moluccanum.
  ,,   genuinum.
Coix Lachryma.
Colubrina asiatica.
Colocasia antiquorum.
  ,,   esculenta.
Commelyna pacifica.
Commersonia platyphylla.
  ,,   echinata.
Connarus Pickeringii.*
  ,,   sp. n. (190).*
  ,,   sp. n.*
Coprosma persicæfolia.*
  ,,   sp. n. (129)*
  ,,   sp. n.*
Cordia subcordata.
  ,,   aspera.
Cordyline Jacquini.
  ,,   sepiaria.
  ,,   terminalia.
Coriaria ruscifolia.
Corymbis veratrifolia.
  ,,   sp. n. (436).*
Couthovia corynocarpa.*
  ,,   sp. n. (924).*
Crinum asiaticum.
Crossostylis Harveyi.*
Crotalaria quinquifolia.
Croton metallicus.*
  ,,   heterotrichus.*
  ,,   Verreauxii.
  ,,   ,,   Storckii.
  ,,   leptopus.*
  ,,   sp. n. (673)*.
Cryptocarya sp. n. (128).*
  ,,   sp. n. (170).*
  ,,   sp. n. (171).*
  ,,   sp. n. (434).*
  ,,   sp. n. (650).*
  ,,   sp. n. (1,068).*

Cryptocarya, sp. n. (1,117).*
(?) Cucurbita Pepo.
(?) Cucumis melo.
(?) „ sativus.
(?) „ acidus.
Cupania rhoifolia.*
„ leptobotrys.*
„ Brackenridgei.*
„ sp. n. (982).*
„ sp. n. (1,000).*
Curcuma longa.
Cuscuaria spuria.*
Cyathula prostrata.
„ debilis.
Cycas circinalis.
Cynometra grandiflora.*
„ falcata.*
„ sp. n. (204).*
„ sp. n. (519).*
„ sp. n.*
Cyperus pinnatus.
„ strigosus.
Cypholophus macrocephalus.
„ heterophyllus.
Cyrtandra Pritchardii.*
„ coloides.*
„ ciliata.*
„ Denhami.*
„ acutangla.*
„ viticnsis.*
„ anthropophago-
      rum.*
„ Milnei.*
„ dolichocarpa.*
„ involucrata.*
„ Harveyi.*
„ Spowarti, N.S.*
„ Butti, N.S.*
„ Chippendalei, N.S.
„ Des-Vœuxi, N.S.
„ Gorriei, N.S.*
„ Langtoni, N.S.*
„ Tempesti, N.S.*
„ sp. n. (439).*

Cyrtandra, sp. n. (578).*
„ sp. n. (698).*
„ sp. n. (1,134).*
Cyrtosperma edulis.*
Dacrydium elatum.
Dalbergia monosperma.
„ sp. n. (509).*
Dammara viticnsis.*
Datura stramonium.
Dichopsis Hornei, N.S.*
„ sp. n. (1,117).*
Dendrobium biflorum.
„ catillare.*
„ mohlianum.
„ crispatum.
„ Gordoni, N.S.*
„ Hornei, N.S.*
„ Tokai.*
„ sp. n. (793).*
„ sp. n. (1,085).*
Desmodium umbellatum.
„ polycarpum.
„ scorpiurus.
Derris uliginosa.
„ sp. n. (1,145).*
Dianella ensifolia.
„ intermedia.
Dioclea violacea.
Dichrocephala latifolium.
Digitaria sanguinalis.
Dioscorea alata.
„ aculeata.
„ bulbiflora.
„ nummularia.
„ pentaphylla.
„ sativa.
Diospyros, sp. n. (195).*
„ sp. n. (904).*
Disemma Barclayi.*
„ Storckii.*
„ viticnsis.*
Dodonæa viscosa.
Dolicholobium Knollysi, N.S.*
„ latifolium.*

Dolicholobium, longissimum.*
„ McGregori, N.S.*
„ oblongifolium.*
„ sp. n. (518).*
„ sp. n. (647).*
„ sp. n. (834).*
„ sp. n.*
Dracontomelon sylvestre.
„ pillosum.* .
Drymispernum Burnettianum.
„ accuminatum.
„ lanceolatum.*
„ montanum.*
„ pubiflorum.*
„ subcordatum.*
Dysoxylum alliaceum.
„ bijugum.
„ sp. n. (131).*
„ sp. n. (141).*
„ sp. n. (316).*
„ sp. n. (375).*
„ sp. n. (410).*
„ sp. n.*
Earina, sp. n. (892).*
Ebermaiera, sp. n. (1,067).*
Eclipta prostrata.
Elatostema nemorosum.*
„ macrophyllum.
„ sp. n. (69).*
„ sp. n. (991).*
Eleocharis articulata.
„ variegata.
Eleusine indica.
Elæocarpus cassinoides.*
„ Græffei.*
„ laurifolius.*
„ pyriformis.*
„ Milnei.*
„ Storckii.*
Endiandra, sp. n. (199).*
Entada scandens.
Epiphanes micradenia.*

Eranthemum laxiflorum.*
„ insularum.*
„ sp. n. (61).*
„ sp. n. (238).*
„ sp. n. (263).*
„ sp. n. (724).*
„ sp. n. (725).*
Eria stenostachya.*
„ sphærocarpa.*
„ æridostachya.
„ rostrifolia.*
„ sp. (116).
Erigeron albidum.
Erythræa australis.
Erythrina ovalifolia.
„ indica.
„ „ alba.
Erythrospermum, sp. n. (79).*
„ sp. n. (150).*
„ sp. n. (712).*
„ sp. n.(1,052).*
Eugenia amicorum.
„ Brackenridgei.*
„ confertiflora.*
„ corynocarpa.
„ gracilipes.*
„ Grayi.*
„ effusa.*
„ neurocalyx.*
„ quadrangulata.*
„ rariflora.
„ rivularis.*
„ rubescens.
„ Richii.
„ mallaccensis.
„ sp. n. (340).*
„ sp. n. (383).*
„ sp. n. (443).*
„ sp. n. (689).*
„ sp. n. (702).*
„ sp. n. (774).*
„ sp. n. (828).*
„ sp. n. (843).*
„ sp. n. (867).*

Eugenia, sp. n. (873).*
," sp. n. (874).*
," sp. n. (914).*
," sp. n. (920).*
," sp. n. (959).*
," sp. n. (1,010).*
," sp. n. (1,031).*
," sp. n. (1,056.)*
," sp. n. (1,100.)*
," sp. n. (1,106.)*
," sp. n.*
," sp. n.*
," sp. n.*
Eulalia japonica.
Euphorbia atota.
" chamissonis.
" fidjiana.*
" pilulifera.
" tanensis.
Eurya angustiflora.
" vitiensis.*
Euxolus caudatus.
Evodia drupacea.
" hortensis.
" Roxburghiana.
" sp. n. (146).*
" sp. n. (975).*
Evolvulus alsinoides.
Excœcaria Agallocha.
Fagræa Berteriana.
" gracilipes.*
" Seemannii.
Faradaya ovalifolia.*
" vitiensis.
" sp. n.*
Ficus aspera.
" Barclayi.*
" Bennettii.
" Cavei, N.S.*
" bambusœfolia.*
" Harveyi.
" Masoni, N.S.*
" Pritchardii.*
" scabra.

Ficus Smithi, N.S.*
" Storckii.*
" theophrastoides.*
" tinctoria.
" vitiensis.*
" obliqua.
" sp. n. (440).*
Fimbristylis communis.
" arvensis.
" dichotoma.
(?) Flacourtia Ramontchi.
Flagellaria elegans.
" indica.
Fleurya interrupta.
Freycinetia vitiensis.*
" Pritchardii.*
" Storckii.*
" Milnei.*
" sp. n. (529).*
" sp. n. (592).*
" sp. n. (844).*
" sp. n. (903).*
Gahnia aspera.
Garcinia sessilis.
" pseudoguttifera.*
" vitiensis.*
" sp. n. (185).*
" sp. n. (450).*
" sp. n. (630).*
" sp. n. (734).*
" taitensis.
" pentagonioides.
" Arthurgordoni, N.S.*
Gardeni Hilli, N.S.*
" Gorriei, N.S.*
" Greivei, N.S.*
" s.p. n. (15).*
" s.p. n. (403).*
Geissosis ternata.*
" " var; n.*
(580)*.

Geitonoplesium cymosum.
„ var. angustifolium.
Geniostoma rupestre.
„ microphyllum.*
Geophila reniformes.
Gironnieria celtidifolia.
Glossogyne tenuifolia.
Glycine tabacina.
Gnetum Gnemon.
(?) Gossypium peruvianum.
(?) „ barbadense.
(?) „ arboreum.
(?) „ tomentosum.
Gouania Richii.*
„ denticulata.
Græffea calyculata.*
Grewia mallococca.
„ persicæfolia.*
„ prunifolia.*
Guettarda speciosa.
„ inconspicua.
„ vitiensis.*
Gymnema stenophyllum.*
„ subnudum.*
Gymnosporia vitiensis.
Gyrocarpus Jacquini.
Habenaria tradescantifolia.*
„ superflua.*
„ supervacanea.*
Haplopetalum Richii.*
„ Seemannii.*
„ s.p. n. (968).*
Hardwickia, s.p. n. (483).*
„ s.p. n. (1,121).*
Hedycarya dorstenioides.*
Hedyotis cratæogonum.
Heliconia, s.p. n.*
Hernandia peltata.
„ s.p. n. (517).*
„ s.p. n. (738).*
Heriteria littoralis.

Hibbertia, s.p. n. (651).*
Hibiscus Rosa-sinensis.
„ Storckii.*
„ diversifollius.
„ esculentus.
„ Abelmoschus.
„ tricuspis.
„ tiliaceus.
„ s.p. n. (704).*
Hiptage myrtifolia.*
„ javanica.
Homalanthus populifolius.
Homalium vitiensis.
Hoya bicarinata.
„ diptera.*
„ Barracki, N.S.*
Hydnophytum (Myrmecodia) imberbe.*
„ longiflorum.*
„ Wilkinsoni, N.S.*
„ Wilsoni, N.S.*
Hydrocotyle asiatica.
Hypolytrum latifolium.
Ilex vitiensis.*
Imperata arundinacea.
Indigofera Anil.
Inocarpus edulis.
Ipomæa pes-capræ.
„ peltata.
„ Terpethum.
„ denticulatum.
Ixora vitiensis.*
„ pelagica.*
„ maxima.*
„ Storckii.*
„ Carewi, N.S.*
„ Joskei N.S.*
Jasminum tetraquetrum.
„ australe
„ didymum.
„ simplicifolium
„ sp. n. (98).*

Jasminum, s.p. n. (672).*
(?) Jatropha curcas.
Karivia vitiensis.
Kleinhovia hospita.
Kentia exorrhiza.*
Kyllinga monocephala.
Lablab vulgaris.
(?) Lagenaria vulgaris.
Lagenophora Pickeringi.*
Laportea Harveyi.*
„ Milnei.*
„ vitiensis.*
Lasianthera vitiensis.*
Lauraceæ, s.p. n. (777).*
Leea sambucina.
Lemna minor.
„ melanorrhiza.
Lepironia mucronata.
Lerchea calycina.*
Litsea (Tetranthera) palma-
tinervia.*
„ vitiense.*
„ Seemannii.*
„ Pickeringii.*
„ s.p. n. (544).*
„ s.p. n. (969).*
„ s.p. n. (733).*
Leucas decemdentata.
Leucoskye corymbulosa
Leucæna Fosteri.
„ glauca.
Leucopogon cymbula.*
Limnophila fragrans.*
Limnanthemum Kleinianum.
Lindenia vitiensis.
Liparis longipes.
„ s.p. n.*
Loranthus insularum.
„ vitiensis.*
„ s.p. n.*
Luffa insularum.
Lumnitzera coccinea.
Lyonsia lævis.*

Maba foliosa.*
„ elliptica.
„ buxifolia.
„ lateriflora, N.S.*
„ s.p. n. (201).*
„ s.p. n. (473).*
„ s.p. n. (823).*
„ s.p. n. (1,050).*
„ s.p. n.*
Macaranga secunda.*
„ Harveyana.*
„ macrophylla.*
„ Maudsleyi, N.S.*
„ membranacea.*
„ Seemannii.
„ s.p. n. (1,044).*
„ s.p. n.*
Mæsa Pickeringii.
„ persicæfolia.
„ nemoralis.
„ vitiensis.*
„ corylifolia.*
„ „ var. n. (822).*
„ s.p. n. (664).*
Malaxis glandulosa.
Mallotus tiliæfolius.
„ s.p. n. (597).*
(?) Mamordica Charantia.
(?) Manihot palmata.
(?) „ „ Aipi.
Maoutia australis.
Mariscus flavus.
„ phleoides.
„ umbellatus.
Marlea vitiensis.
„ „ var. n. (408).*
Medinilla heterophylla.*
„ rhodochlæna.*
„ amœna.
„ Waterhousei.*
„ parvifolia.*
„ s.p. n. (182).*
„ s.p. n. (944).*
Melastoma denticulata.

·Melastoma Novæ Hollandiæ.
(?) Melia elegans.
Melochia odorata.
„ vitiensis.
Melodinus scandens.
Melodorum, s.p. n. (318).*
Memecylon vitiense.*
Metrosideros polymorpha.
„ „ var. fls. yellow (764).
„ „ var. fls. white (933).
Micromelum minutum.
Microstylis purpurea.
„ platychila.*
(?) Mimosa pudica.
Missiessya corymbulosa.
Mollugo stricta.
Monosis insularum.
Morinda citrifolia.
„ Fosteri.
„ myrtifolia.*
„ Grayi.*
„ mollis.
„ bucidæfolia*.
„ s.p. n. (347).*
„ s.p. n. (1,128).*
(?) Morus indica.
Mucuna platyphylla.
„ gigantea
„ sp. n. (721).*
Musa sapientum.
„ cavendishii.
„ paradisiaca.
„ uranocarpus.
Mussænda frondosa.
Myriogyne minuta.
Myristica castanæfolia.*
„ grandifolia.*
„ sp. n. (243).*
„ sp. n. (966).*
Myrsine crassifolia.
„ myricæfolia.
„ Brackenridgei.*

Myrsine, sp. n. (261).*
„ sp. n. (291).*
„ sp. n. (355).*
„ sp. n. (455).*
Myrtaceæ, sp. n.
„ (?) sp. n. (878).*
Nauclea Fosteri.
Nephelium (Pometia) pinnatum.
Nesopanax vitiensis.*
Nelitris vitiensis.
„ fruticosa.
? Nicotiana Tabacum.
Nothopanax fruticosum.
„ multijugum.*
Ochrosia parviflorum.
Ocymum gratissimum.
Oldenlandia tenuifolia.
„ paniculata.
„ Burmanniana.
„ sp. n.*
Olea vitiensis.*
Olyra micrantha.
Oncocarpus vitiensis.*
„ sp. n. (629).*
Ophiorrhiza peploides.*
„ leptantha.*
„ laxa.*
„ „ var.? (213).*
Oplismenus Burmanni.
„ pupurascens.
„ albo-striatus.
„ compositus.
Orchideæ sp. n. (115).*
Ormocarpum sennoides.
Orthoclada sp. n. (107).*
Oxalis corniculata.
Pachyrrhizus trilobus.
Pandanus caricosus.
„ virens.
„ odoratissimus.
„ Joskei, N.S.*
Panicum sanguinale.
„ trigonum.

Panicum, ambiguum.
Papaya (Carica papaya) vulgaris.
Paphia vitiensis.*
Parinarium insularum.
„ laurinum.
Parkia Pari, N.S.*
Paspalum scrobiculatum.
Payena Hilli, N.S.*
Pelagodendron vitiensis.*
Pellionia elatostemoides.
„ „ minor.
„ filicoides.*
„ australis.*
„ vitiensis.
„ sp. n. (817).*
Peperomia pallida.
„ sp. n. (30).*
„ sp. n. (370).*
„ sp. n. (652).*
Peristylis sp. n. (448).*
Phajus Blumei.
Pharbitis insularis.
Phaseolus truxillensis.
„ mungo.
Phyllanthus ramiflorus.
„ „ lanceolatus.
„ „ var. n. (1,009).*
„ cordatus.*
„ concolor.
„ „ ellipticus.
„ „ ohonatus.
„ pacificum.*
„ vitiensis.
„ Seemannianus.*
„ venulosis.*
„ cordatus.
„ manono.
„ pedocarpus.
„ amentuliger.*
„ heterodoxus.*
„ simplex.
„ Wilkesianus.*

Phyllanthus, sp. n. (120a).*
„ sp. n. (364).*
„ sp. n. (676).*
„ sp. n. (767.)*
„ sp. n. (1,120).*
„ sp. n. (1,125).*
(?) Physalis peruviana.
„ angulata.
Pimia rhamnoides.*
Piper methysticum.
„ latifolium.
„ Macgillivrayi.
„ insectifugum.*
„ sp. n. (48).*
„ sp. n. (839).*
„ sp. n.*
Pipturus velutinus.
„ propinqua.
„ argenteus.
„ platyphyllus.*
„ sp. n. (72).*
Pisonia inermis.
„ umbellifera
Pittosporum Brackenridgei.*
„ Pickeringii.*
„ rhytidocarpum.*
„ arborescens.
„ Richii.*
„ tobiroides.*
„ sp. n. (667).*
„ sp. n. (989).*
„ sp. n. (1,084).*
Plantago major.
Plectranthus Forsteri.
Plectronia McGregori, N.S.*
„ McConnelli, N.S.*
„ sp. n. (261).*
„ sp. n. (614).*
Plerandra Grayi.*
„ Pickeringii.
Plumbago zeylanica.
Podocarpus affinis.*
„ bracteata.
„ cupressina.

Podocarpus vitiensis.*
Pogonia sp. n.*
Polyalthia vitiensis.
Polyglonum glabrum.
Pongamia glabra.
Portulacca quadrifida.
„ oleracea.
Premna taitensis.
Pritchardi pacifica.
Procris cephalida.
Psychotria Brackenridgei.*
„ Broweri.*
„ cordata.*
„ calycosa.*
„ bullata.*
„ Fosteriana.*
„ filipes.*
„ gracilis.*
„ hyporgyraea.*
„ macrocalyx.
„ pelagica.*
„ Pritchardii.*
„ platycocca.*
„ serpens.
„ sulphurea.*
„ Storckii.*
„ tephrosantha.*
„ tetragona.*
„ turbinata.*
„ Pickeringii.*
„ sp. n. (31).*
„ sp. n. (33.)*
„ sp. n. (71).*
„ sp. n. (78).*
„ sp. n. (83).*
„ sp. n. (173).*
„ sp. n. (292).*
„ sp. n. (587a).*
„ sp. n. (575).*
„ sp. n. (586).*
„ sp. n. (691.*
„ sp. n. (708).*
„ sp. n. (790).*
„ sp. n. (831).*

Psychotria, sp. n. (849).
„ sp. n. (994a).*
„ sp. n. (1,042).*
„ sp. n. (1,115a).*
? Pterocarpus indicus.
Pterospermum sp. n.*
Ptychosperma filiferum.*
„ pauciflorum.*
„ perbreve.*
„ Pickeringii.*
„ Seemannii.*
„ vitiensis.*
Ratonia falcata.*
„ sapindus.
„ Storckii.*
Rhamnus vitiensis.
„ sp. n. (1,115).*
„ sp. n. (1,116).*
Rhamphidia rubicunda.
Rhaphidophra vitiensis.*
„ Storckiana.*
Rhizophora mucronata.
Rhus simarubaefolia.*
„ taitensis.
Rhynchospora aurea.
Rhytidandra vitiensis.
Richella monosperma.
(?) Ricinus communis.
Rourea heterophylla.*
„ sp. n. (349).*
„ sp. n. (394).*
Rubiaceae sp. n. (441).*
„ sp. n. (571).*
„ sp. n. (608).*
„ sp. n. (986).*
„ sp. n. (1,132).*
Rubus panicullatus.
„ tiliaceus.
Saccharum officinarum.
„ sp. n. (203).*
Saccolabium Bertholdi.*
Sagus vitiensis.
Salacia sp. n. (791).*
„ sp. n. (1,127).*

Santalum (yasi) album.
Sapindus vitiensis.*
Sapota pyrulifera.*
„ vitiensis.*
„ sp. n.*
Sapotaceæ sp. n. (95).*
„ sp. n. (187).*
„ sp. n. (477).*
„ sp. n. (827).*
„ sp. n. (829).*
Saracanthus sp. n. (886).*
„ nagarensis.*
Sauranja rubicunda,*
„ grandiflora.
„ var. n. (907).*
„ var. n. (934).*
„ sp. n. (847).*
Schefileria vitiense.*
Schmidelia glabra.
Schizostachyum glaucifolium.
Scleria margaritifera.
„ lethosperma.
Scævola floribunda.*
„ Koenigii.
„ sericea.
Scrianthes myriadenia.
„ vitiensis.*
Sida linifolia.
„ microphylla.
„ rhombifolia.
Sideroxylon sp. n. (317).*
„ sp. n. (317a).*
„ sp. n. (1,140).*
„ sp. n.*
Siegesbeckia orientalis.
Smilax trifurcata.
„ (Pleiosmilax) vitiensis.*
„ sp. n. (634).*
Symthea Pacifica.*
Solanum oleraceum.
„ anthropophagorum.
„ repandum.
„ tetandrum.
„ tuberosum.

Salanum vitiensis.
„ Seedi, N.S.*
„ sp. n. (595).*
„ sp. n. (678).*
„ sp. n. (679).*
„ sp. n. (714).*
Sophora tomentosa.
Sonchus aspera.
Spathoglottis pacifica.
Spiræanthemum Græffei.*
„ katakata.*
„ samoense.
„ vitiense.*
Spondias dulcis.
Sponia velutina.
„ Andersonii.
Stemonurus vitiensis.
Sterculia diversifolia.*
„ vitiensis.*
„ sp. n. (357).*
Stillingia pacifica.
Storckiella vitiensis.*
Strongylodon glaber.
„ lucidum.
Strychnos colubrina.
Stylocoryne coffæoides.
„ Harveyi.*
„ sambucina.
Symplocos spicata.
„ sp. n. (54).*
„ sp. n. (175).*
Tabernæmontana orientalis.
„ pacifica.*
„ Thurstoni, N.S.
„ sp. n. (58)*
„ sp. n.*
Tacca pinnatifida.
„ maculata.
„ sp. n. (598).*
Tænoiphyllum Seemannii.*
Talinum patens.
Teucrum inflatum.
Tephrosia piscatoria
„ purpurea.

Terminalia catappa.
„ litoralis.
„ sp. n. (420).*
„ sp. n. (488).*
Ternstræmia vitiensis.*
„ sp. n. (930).*
Thacombauia vitiensis.*
Thespesia populnea.
Thouarea involuta.
Thrixpermum Godeffroy-
   anum.*
Timonius affinis.*
„ sapotæfolius.*
Tournefortia argentea.
Trichospermum Richii*
Trimenia weimannifolia.*
Triumfetta procumbens.
Tropidia effusa.*
Trophis anthropophagorum.*
Tylophora Brackenridgei.*
Typha angustifolia.
Unona sp. n. (287).*
Uvaria amygdalina.*
Uraria lagopoides.
Urena lobata.
„ morifolia.
Vandellia crustacea.
Vavæa Harveyi.*
„ vitiensis.*
Veitchia Storckii.*
„ joannis.*
„ subglobosa.*

Ventilago vitiensis.*
Viguea lutea.
Viscum articulatum.
„ sp. n. (894).*
Vitex acuminata.
„ trifolia.
„ vitiensis.*
„ sp. n.*
Vitis saponaria.
„ vitiensis.
„ acuminata.
Vrydagzynea purpurea.
Waltheria americana.
Weinmannia affinis.*
„ Richii.*
„ spiræoides.
„ vitiensis.*
„ sp. n. (632).*
Wickstræmia foetida.
Wollastonia strigulosa.
Wormia biflora.*
„ membranifolia.*
Ximenia elliptica.
Xylosma orbiculata.
„ sp. n. (301).*
(?) Zea mays.
Zingiber Zerumbet.
Species indetermined (1,121).*
„ „ (118).*
„ „ (395).*
„ „ (600).*
„ „ No. Nr.*

SYSTEMATIC LIST OF NATURAL ORDERS and genera to which the plants found in Fiji belong, as arranged in Seemann's Flora Vitiensis. Endemic genera are marked by an asterisk (*) before them; the figures after them indicate the number of species in each genus that have been discovered in Fiji, by myself and previous travellers.

Order I.
Ranunculaceæ.
 Clematis, 1.

II.
Dilleniaceæ.
 Wormia, 2.
 Hibbertia, 1.

III.
Anonaceæ.
 Uvaria, 1.
 Polyalthia, 1.
 Cananga, 1.
 *Richella, 1.
 Unona, 1.
 Melodorum, 1.

IV.
Cruciferæ.
 Cardamine, 1.

V.
Capparideæ.
 Capparis, 1.

VI.
Violareæ.
 Agation (Agatea), 2.
 Alsodeia, 1.

VII.
Bixineæ.
 Xylosma, 2.
 Bixa, 1.

Flacourtia, 1.
Erythrospermum, 4.

VIII.
Pittosporeæ.
 Pittosporum, 9.

IX.
Portulaceæ.
 Portulaca, 2.
 Talinum, 1.

X.
Elatineæ.
 Elatine, 1.

XI.
Gutifferæ.
 Garcinia, 7.
 Calophyllum, 4.
 Calysaccion, 1.

XII.
Ternstroemiaceæ.
 Ternstroemia, 2.
 Eurya, 2.
 Sauranja, 3.
 *Trimenia, 1.

XIII.
Malvaceæ.
 Sida, 3.
 Urena, 2.
 Hibiscus, 8.
 Thespesia, 1.

Gossypium, 4.
Abutilon, 1.

XIV.
Sterculiaceæ.
  Sterculia, 3.
  Heritiera, 1.
  Kleinhovia, 1.
  Melochia, 2.
  Waltheria, 1.
  Commersonia, 2.
  *Pimia, 1.
  Pterospermum, 1.

XV.
Tiliaceæ.
  Grewia, 3.
  Triumfetta, 1.
  *Graeffea, 1.
  Trichospermum, 1.
  Elæocarpus, 6.

XVI.
Malpighiaceæ.
  Hiptage, 2.

XVII.
Geraniaceæ.
  Oxalis, 1.

XVIII.
Rutaceæ.
  Evodia, 5.
  Acronychia, 2.
  Micromelum, 1.
  Citrus, 4.

XIX.
Simarubeæ.
  Brucea, 1.
  *Amaroria, 1.

XX.
Ochnaceæ.
  Brackenridgea, 4.

XXI.
Bruseraceæ.
  Canarium 3.

XXII.
Meliaceæ.
  Vavæa, 2.
  Melia, 1.
  Dysoxylum, 8.
  Aglaia, 6.
  Carapa, 2.

XXIII.
Chailletiaceæ.
  Chailletia, 1.

XXIV.
Olacineæ.
  Ximenia, 1.

XXV.
Icacineæ.
  Stemonurus, 1.

XXVa.
Humiriaceæ.
  *Thacombauia, 1.

XXVI.
Illicineæ.
  Ilex, 1.

XXVII.
Celastrineæ.
  Celastrus, 1.
  Gymnosporia, 1.
  Salacia, 2.

XXVIII.
Rhamneæ.
  Ventilago, 1.
  *Smythea, 1.
  Rhamnus, 3.

Colubrina, 1.
Alphitonia, 1.
Gouania, 2.

XXIX.
Ampelideæ.
Vitis, 3.
Leea, 1.

XXX.
Sapindaceæ.
Cardiospermum, 2.
Cupania, 5.
Ratonia, 3.
Sapindus, 1.
Pometia (Nephilium), 1.
Dodonæa, 1.
Schmidelia, 1.

XXXI.
Anacardiaceæ.
Rhus, 2.
Buchanania, 1.
*Oncocarpus, 2.
Spondias, 2.
Dracontomelon, 2.
Anacardium, s.p. n., 1.

XXXII.
Connaraceæ.
Rourea, 3.
Connarus, 3.

XXXIII.
Leguminosæ.
Crotalaria, 1.
Indigofera, 1.
Tephrosia, 2.
Ormocarpum, 1.
Desmodium, 3.
Uraria, 1.
Glycine, 1.
Dioclea, 1.
Canavalia, 3.

Mucuna, 3.
Erythrina, 2.
Strongylodon, 2.
Phaseolus, 2.
Vigna, 1.
Lablab, 1.
Pachyrrhizus, 1.
Abrus, 1.
Pterocarpus, 1.
Dalbergia, 2.
Pongamia, 1.
Derris, 2.
Sophora, 1.
Cæsalpinia, 2.
Cassia, 6.
Storckiella, 1.
Afzelia, 1.
Bauhinia, 1.
Inocarpus, 1.
Cynometra, 5.
Entada, 1.
Mimosa, 1.
Leucæna, 2.
Acacia, 2.
Serianthes, 2.
Parkia, 1.
Hardwickia, 2.

XXXIV.
Chrysobalaneæ.
Parinarium, 2.

XXXV.
Rosaceæ.
Rubus, 1.

XXXVI.
Myrtaceæ.
Eugenia (Jambosa), 36.
Nelitris, 2.
Calyptranthes (Acicalyptus), 4.
Barringtonia, 4.
Metrosideros, 1.

XXXVII.
Melastomaceæ.
Memecylon, 1.
Astronia, 8.
Astronidium, 2.
Amplectrum, 1.
Medinilla, 7.
Melastoma, 2.

XXXVIII.
Rhizophoraceæ.
*Haplopetalon, 3.
Rhizophora, 1.
Bruguiera, 1.
Crossostylis, 1.

XXXIX.
Combretaceæ.
Terminalia, 4.
Lumnitzera, 1.
Gyrocarpus, 1.

XL.
Homalineæ.
Homalium, 1.

XLI.
Passifloraceæ.
Disemma, 3.

XLIa.
Begoniaceæ.
Begonia, 1.

XLII.
Papayaceæ.
Papaya (Carica), 1.

XLIII.
Samydeæ.
Casearia, 8.

XLIV.
Balanophoreæ.
Balanophora, 1.

XLV.
Taccaceæ.
Tacca, 3.

XLVI.
Cucurbitaceæ.
Karvivia, 1.
Citrullus, 1.
Mamordica, 1.
Lufa, 1.
Lagenaria, 1.
Cucumus, 3.
Cucurbita, 1.
Alsomitra, 1.

XLVII.
Saxifragaceæ.
Geissois, 1.
Weinmannia, 5.
Spiræanthemum, 4.

XLVIII.
Hederaceæ.
Marlea, 2.
Hydrocotyle, 1.
Nothopanax, 2.
Aglama, 1.
Schefflera, 1.
*Nesopanax, 1.
*Bakeria, 1.
Plerandra, 2.
Arthropanax, 1.

XLIX.
Coraceæ.
Rhytidandra, 1.

L.
Loranthaceæ.
Viscum, 2.
Loranthus, 3.

LI.
Rubiaceæ.
Dolicholobium, 9.

Gardenia, 9.
Mussænda, 1.
Stylocoryne, 3.
*Pelagodendron, 1.
Hedyotis, 1.
Oldenlandia, 4.
Ophiorrhiza, 3.
Lerchea, 1.
Lindenia, 1.
Morinda, 8.
Timonius, 2.
Guettarda, 3.
Calycosia, 14.
Ixora, 6.
Psychotria, 38.
Hydnophytum (Myrmecodia), 4.
Geophila, 1.
Coprosma, 3.
Canthium, 4.
Nauclea, 1.
Badusa, 1.
Plectronia, 4.

LII.
Compositæ.
Monosis, 1.
Ageratum, 1.
Adenostemma, 1.
Erigeron, 1.
Blumea, 2.
Eclipta, 1.
Siegesbeckia, 1.
Wollastonia, 1.
Bideus, 1.
Lagenophora, 1.
Dichrocephala, 1.
Glossogyne, 1.
Myriogyne, 1.
Sonchus, 1.

LII.
Goodeniaceæ.
Scævola, 3.

LIV.
Ericaceæ.
*Paphia, 1.
Leucopogon, 1.

LV.
Myrsineæ.
Mæsa, 7.
Myrsine, 7.
Ardisia, 10.

LVI.
Sapotaceæ.
Sapota, 3.
Sideroxylon, 4.
Payena, 1.
Dichopsis, 2.

LVII.
Ebenaceæ.
Maba, 9.
Diospyros, 2.

LVIII.
Styraceæ.
Symplocos, 3.

LIX.
Jasmineæ.
Jasminium, 6.
Olea, 1.

LX.
Apocyneæ.
Melodinus, 1.
*Carruthersia, 3.
Alyxia, 4.
Cerbera, 1.
Ochrosia, 1.
Tabernæmontana, 5.
Lyonsia, 1.
Alstonia, 6.

LXI.
Asclepiadeæ.
　Tylophora, 1.
　Gymnema, 2.
　Hoya, 3.

LXII.
Loganiaceæ.
　Geniostoma, 2.
　Fagræa, 3.
　*Couthovia, 2.
　Strychnos, 1.
　*Canthiopsis, 1.

LXIII.
Gentianeæ.
　Erythræa, 1.
　Limnanthemum, 1.

*LXIV.
Boragineæ.
　Cordia, 2.
　Tournefortia, 1.

LXV.
Convolvulaceæ.
　Batatas, 2.
　Pharbitis, 1.
　Calonyction, 2.
　Ipomæa, 4.
　Aniseia, 2.
　Evolvulus, 1.

LXVI.
Solanaceæ.
　Solanum, 11.
　Capsicum, 1.
　Physalis, 2.
　Datura, 1.
　Nicotiana, 1.

LXVII.
Scrophularineæ.
　Limnophila, 1.
　Vandellia, 1.

LXVIII.
Cyrtandreæ.
　Cyrtandra, 22.

LXIX.
Acanthaceæ.
　Adenosma, 1.
　Chæctacanthus, 1.
　Eranthemum, 7.
　Ebermaiera, 1.

LXX.
Verbenaceæ.
　Premna, 1.
　Clerodendron, 4.
　Faradaya, 3.
　Vitex, 4.

LXXI.
Labiatæ.
　Ocymum, 1.
　Plectranthus, 1.
　Leucas, 1.
　Teurcrium, 1.

LXXII.
Plantagineæ.
　Plantago, 1.

LXXIII.
Plumbagineæ.
　Plumbago, 1.

LXXIV.
Nyctagineæ.
　Pisonia, 2.
　Boerhaavia, 2.

LXXV.
Amarantaceæ.
　Amaranthus, 4.
　Euxolus, 1.
　Achryanthes, 1.
　Cyathula, 2.

S 2

LXXVI.
Molluginaceæ.
　Mollugo, 1.

LXXVII.
Polygonaceæ.
　Polygonum, 1.

LXXVIII.
Lauraceæ.
　Cinnamomum, 7.
　Litsea (Tetranthera), 7.
　Cassytha, 1.
　Cryptocarya, 5.
　Endiandra, 1.

LXXIX.
Hernandiaceæ.
　Hernandia, 3.

LXXX.
Myristicaceæ.
　Myristica, 4.

LXXXI.
Monimiaceæ.
　Hedycarya, 1.

LXXXII.
Thymelæaceæ.
　Wickstræmia, 1.
　Drymispermum, 6.

LXXXIIa.
Aristolochiaceæ.
　Aristolochia sps. n. 1.

LXXXIII.
Santalaceæ.
　Satanlum, 1.

LXXXIV.
Euphorbiaceæ.
　Euphorbia, 5.
　Antidesma, 3.
　Phyllanthus, 21.
　Baccaurea, 6.
　Bischoffia, 1.
　Croton, 5.
　Aleurites, 1.
　Claoxylon, 2.
　Acalypha, 12.
　Mallotus, 2.
　Cleidion, 2.
　Macaranga, 8.
　Ricinus, 1.
　Manihot, 1.
　Jatropha, 1.
　Codiæum, 4.
　Carumbium, 1.
　Stillingia, 1.
　Excæcaria, 1.
　Homalanthus, 1.

LXXXV.
Urticaceæ.
　Sponia, 2.
　Grionnieria, 1.
　Fleurya, 1.
　Laportea, 3.
　Pellionia, 4.
　Elatostemma, 4.
　Proceris, 1.
　Boehmeria, 3.
　Cypholophus, 2.
　Pipturus, 5.
　Missiessya, 1.
　Maoutia, 1.
　Morus, 1.
　Broussonetia, 1.
　Ficus, 16.
　Antiaris, 1.
　Caturus, 1.
　Artocarpus, 1.
　Trophis, 1.
　Leucoskye, 1.

LXXXVa.
Celtideæ.
    Celtis, 1.

LXXXVI.
Ceratophylleæ.
    Ceratophyllum, 1.

LXXXVII.
Chloranthaceæ.
    Ascarina, 1.

LXXXVIII.
Piperaceæ.
    Peperomia, 4.
    Piper, 7.

LXXXIX.
Casuarineæ.
    Casuarina, 2.

XC.
Coniferæ.
    Dammara, 1.
    Podocarpus, 4.
    Dacrydium, 1.
    Gnetum, 1.

XCI.
Cycadeæ.
    Cycas, 1.

XCII.
Palmaceæ
    Kentia, 1.
    Veitchia, 3.
    Ptychospermum, 6.
    Pritchardia, 1.
    Calamus, 1.
    Cocos, 1.
    Sagus, 1.

XCIII.
Pandanaceæ.
    Typha, 1.

    Pandanus, 4.
    Freycinetia, 8.

XCIV.
Aroideæ.
    Amorphophallus, 1.
    Colocasia, 2.
    Alocasia, 1.
    Rhaphidophora, 2.
    Cuscuaria, 1.
    Cyrtosperma, 1.

XCV.
Lemnaceæ.
    Lemna, 2.

XCVI.
Scitamineæ.
    Heliconia, 1.
    Musa, 4.
    Alpinia, 6.
    Amomum, 2.
    Curcuma, 1.
    Zingiber, 1.
    Canna, 1.

XCVII.
Orchideæ.
    Habenaria, 3.
    Vrydagzynea, 1.
    Anæctochilus (Anecochilus), 1.
    Rhamphidia, 1.
    Tropidia, 1.
    Corymbis, 2.
    Epiphanes, 1.
    Pogonia, 1.
    Tænophyllum, 1.
    Thrixspermum, 1.
    Saccolabium, 2.
    Sarcanthus, 2.
    Calanthe, 4.
    Appendiculata, 2.
    Phajus, 1.
    Spathaglottis, 1.

Eria, 5.
Liparis, 2.
Malaxis, 1
Microstylis, 2.
Bulbophyllum, 2.
Dendrobium, 9.
Chrysoglossum, 1.
Earina, 1.
Peristylis, 1.

XCVII.
Amaryllideæ.
Crinum, 1.

XCIX.
Dioscoreæ.
Dioscorea, 6.

C.
Smilaceæ.
Smilax, 3.

CI.
Liliaceæ.
Cordyline, 3.
Geitonoplesium, 2.
Dianella, 2.
Astelia, 2.

CII.
Commelynaceæ.
Commelyna, 1.
Aneilema, 1.

CIII.
Juncaceæ.
Flagellaria, 2.

CIV.
Cyperaceæ.
Lepironia, 1.
Scleria, 2.
Gahnia, 1.
Rhynchospora, 1.
Hypolytrum, 1.
Fimbristylis, 3.
Eleocharis, 2.
Kyllingia, 1.
Mariscus, 3.
Cyperus, 2.

CV.
Gramineæ.
Andropogon, 2.
Eulalia, 1.
Saccharum, 2.
Imperata, 1.
Eleusine, 1.
Centotheca, 1.
Schizostachyum, 1.
Thouarea, 1.
Cenchrus, 1.
Oplismenus, 4.
Panicum, 3.
Paspalum, 1.
Olyra, 1.
Coix, 1.
Zea, 1.
Chloris, 1.
Orthoclada, 1.

CVI.

FILICES AND ALLIED ORDERS.

Note.—Species marked with an asterisk (*) indicate that they are new and endemic. Species marked (*o*) indicates that they are new to Polynesia.

Gleichenia oceanica, Kuhn.
„ dichotoma, Hook.
„ flagellaris, Spreng.
Cyathea propinqua, Mett.
Alsophila lunulata, R. Br.
* „ Hornei, Baker.
„ truncata, Brack.
o Hymenophyllum javanicum, Spreng.
„ dilatum, Sw.
„ tunbridgense, Sw.
„ multifidum, Sw.
„ flabellatum, Labell.
„ Neesi, Hook.
* Trichomanes cultratum, Baker.
„ muscoides, Sw.
„ proliferum, Blume.
„ javanicum, Blume.
„ pyxidiferum, L.
„ caudatum, Brack.
„ maximum, Blume.
„ rigidum, Sw.
„ humile, Forst.
„ saxifragoides, Presl.
„ alternans, Carr.
„ vitiense, Baker.
„ fœniculatum, Bory.
„ apiifolium, Presl.
Dicksonia Brackenridgei, Mett.
„ straminea, Labell.
* „ moluccana, var. inermis, Baker.
* „ incurvata, Baker.
Davallia contigua, Sw.
„ heterophylla, Sw.
„ polypodioides, Brack.
„ botrychioides, Brack.
„ pentaphylla, Blume.

Davallia fijiensis, Hook, (Endemic).
„ pinnata, Carr.
„ rhomboidea, Wall.
„ strigosa, Sw.
„ Denhamii, Hook, (Endemic).
„ solida, Sw.
„ epiphylla, Sw.
„ tenuifolia, Sw.
„ Laperousii, Hook.
„ moluccana, Blume.
„ gibberosa, Sw.
„ fœniculacea, Hook.
„ ferulacea, Moore, (Endemic).
o „ hymenophylloides, Baker.
„ Blumeana, Hook.
„ pulchella, Hook.
„ stolonifera, Baker.
„ „ var. acutifolia.
„ repens, Deso.
„ lobata, Poir.
„ „ „ var. Harveyi, Carr.
„ „ „ var. Seemannii, Carr.
„ ensifola, Sw.
„ parallea, Wall.
„ alpina, Blume.
„ speluncæ, Baker.
„ tenuis, Brack, (Endemic).
Hypolepis tenuifolia, Bernh.
* Adiantum Hornei, Baker.
„ lunulatum, Burm.
„ diaphanum, Blume.
„ hispidulum, Sw.
„ fulvum, Raoul.
o Cheilanthes farinosa, Kaulf.
„ tenuifolia, Sw.
Notholæna hirsuta, Desv.
Doodia media, R. Br.
Pillæa geraniæfolia.
Pteris quadriaurita, Retz.
„ aquilina var. esculenta, Forst.
„ patens, Hook.
„ ensiformis, Burm.
* „ viticnsis, Baker.
„ longifolia, L.
„ incisa, Thunb.

Pteris comans, Forst.
„ Milnei, Baker.
„ tripartita.
Blechnum orientale, L.
Lomaria Pattersoni, Spreng.
„ elongata, Blume.
„ attenuata, Wild.
„ lanceolata, Spreng.
„ procera, Spreng.
„ adnata, Blume.
„ volcanica, Blume.
„ filiformis, Cunn.
Asplenium nidus, L.
„ amboinense, Willd.
* „ „ var. Hilli.
„ multilineatum, Hook.
„ vittæforme, Carr.
„ fijiense, Brack.
„ Carruthersii, Baker.
„ tenerum, Forst.
„ resectum, Sm.
„ falcatum, Lam.
„ caudatum, Forst.
„ cuneatum, Lam.
„ laserpitiifolium, Lam.
„ affine Sw.
. „ bipinnatum, Brack.
„ obtusifolium, Hook.
„ sylvaticum, Presl.
„ japonicum, Thunb.
„ Brackenridgei. Baker.
* „ maximum, Don.
„ melanocaulon, Baker.
„ decussatum, Sw.
„ esculentum, Presl.
„ arborescens, Mett.
„ rhizophyllum, Kunze.
„ multifidum, Brack.
„ polypodioides, Blume.
„ induratum, Hook.
o Allantoida Brunoniana, Wall.
Didymochloena lunulata, Desv.
Aspidium aristatum, Sw.
o „ aculeatum, Sw.

Aspidium semicordatum, Sw.
Nephrodium Prenticii, Baker.
,, albopunctatum, Desv
,, Brackenridgii, Baker.
,, dissectum, Desv.
,, Milnei, Hook.
,, velutinum, Hook.
,, Harveyi, Presl.
,, Luzeanum, Hook.
,, decurrens, Baker.
,, pachyphyllum, Baker.
,, latifolium, Baker.
* ,, tripartitum, Baker.
* ,, juglandifolium, Baker.
,, squamigera, Hook.
,, tenuifolia, Hook.
,, fijiense, Hook.
,, davallioides, Baker.
,, Preslei, Baker.
,, truncatum, Presl.
,, molle, Desv.
Nephrolepis acuta, Presl.
,, ramosa, Moore.
,, cordifolia, Presl.
,, exaltata, Sw.
Oleander mollis, Presl.
* Polypodium deparioides, Baker.
* ,, Gordoni, Baker.
* ,, punctatum, Thunb.
* ,, alsophiloides, Baker.
0 ,, ornatum, Wall.
,, difforme, Blume.
,, rubrinervæ, Baker.
0 ,, cucullatum, Nees.
,, blechnoides, Hook.
* ,, Hornei, Baker.
,, subauriculatum, Blume.
,, adnescens, Sw.
,, irioides, Lam.
,, triquetrum, Blume.
,, linguæforme, Mett.
,, accedens, Blume.
,, ligulatum, Baker.
,, Brownii, Wickst.

Polypodium alatum, Hook.
„ nigrescens, Hook.
„ Phymatodes, L.
* „ viticnse, Baker.
„ rigidulum, Sw.
„ dipteris, Blume.
„ simplicifolium, Hook.
„ Linnæi, Bory.
„ Hookeri, Brack.
„ costatum, Hook.
o Meniscium Beccarianum, Cesati.
Monogramme Junghuhunii, Hook.
Vittaria elongata, Sw.
o „ scolopendrina, Thwaites.
Antrophyum subfalcatum, Brack.
o „ reticulatum, Kaulf.
„ plantagineum, Kaulf.
„ semicostatum, Blume.
Gymnogramme lanceolata, Hook.
„ javanica, Blume.
„ pinnata, Hook.
o „ „ var. brachysora, Baker.
o „ „ var. polypodioides, Baker.
o „ Wallichi, Hook.
* „ scolopendrioides, Baker.
o „ borneensis, var. major.
„ decipens, Mett.
„ caudiformis, Hook.
Hemionitis lanceolata, Hook.
Acrostichum obtusifolium, Brack.
o „ conforme, Sw.
„ scandens, J. Sm.
o „ gorgoneum, Kaulf.
„ sorbifolium, L. var. subtrifoliatum, H.K.
„ „ var Seemannii, Carr.
„ fijiense, Hook.
„ spicatum, Hook.
„ polyphyllum, Hook.
o „ Blumeanum, Hook.
„ „ bipinnatum.
„ cultratum, Baker.
„ repandum, Blume.
„ „ var. Quoyanum, Gand.
„ rivulare, Baker.

Acrostichum aureum, L.
Schizæa dichotoma, Sw.
„ digitata, Sw.
Marattia Fraxinea, Sw.
„ Douglasii?
Lygodium reticulatum, Schk.
Ophiglossum pendulum, L.
„ nudicaule, L.
Todea Wilkesiana, Brack.
Lycopodium cernuum, L.
    „ serratum, Thunb.
    „ squarrosum, Forst.
    „ nummularifolium, Blume.
    „ Phlegmaria, L.
    „ laxum, Presl.
    „ carinatum, Desv.
    „ volubile, Forst.
o Selanginella latifolia, Spreng.
    „ caudata, Spreng.
    „ viridangula, Spreng.
    „ ciliaris, Brack.
    „ flabellata, Spreng.
    „ „ firma.
    „ Wallichii, Spreng.
    „ Menziesii, Spreng.
    „ atrovirides, Spreng.
Psilotum triquetrum, Sw.
   „ complanatum, Sw.
   „ flaccidum, Wall.
Angiopteris evicta, Hoffm.
Equisetum debile, Roxb.
Marsilea, sps.

LIST OF PLANTS found in Upolu, while on a short visit to the Samoan, or Navigator Islands.

FERNS.

Acrostichum fijiense, Hook.
   „ repandum, Blume.
Adiantum lunulatum, Burm.
Asplenium horridum, Raulf.
Antrophyum Grevillei, Balf.

Davallia, Emersoni, H.A.G.
„ heterophylla, Sw.
„ vestita, Blume.
Hymenophyllum dilatum, Sw.
Lomaria volcanica, Blume.
Nephrodium Haenkianum, Presl.
„ heptaphyllum, Baker
Nephrolepis blechnoides, J. Sm.
Polypodium cucullatum, Nees.
Pteris quadriaurita, Retz.
„ patens, Hook.
Ophiglossum reticulatum, L.
Trichomanes, apiifolium, Presl.
„ Powellii.

Flowering Plants.

Alstonia, sp. n.
Barringtonia, sp. n. (314).
Calanthe, sp. n. (24).
Canarium, sp. n. (5).
Clerodendron amicorum.
Crossostylis, sp. n.
Cypholophus, sp. n. (35).
Dendrobium Gordoni, N. S.
Diospyros Samoense.
Dysoxylum sp. n., 7.
„ sp. n. 28.
Elæocarpus Græffi.
Leucosmia, sp. n. (33 and 49).
Lignotidæ, sp. n.
Micromelum minutum.
Morinda, sp. n. (55).
Myristica hypargyrea.
„ N.S. near „ (10).
Orchid, sp. (24), indetermined.
„ sp. (24).
„ sp. (44).
Peperomia, sp. n. (25).
Premna taitense.
Psychotria insularum.
„ sp. n. (43).
Rhamnus, sp. n. (31).

Rubiaceæ, sp. n. (4).
Santalum, sp. n. (13).
Sideroxylon, sp. n. (2).
„ sp. n.
Solanum vitiensis.
Slylocoryne sambucina.
Vrydagzynea.

INDEX.

A.

Acacia richii, 208.
Acclimatisation, of animals, 140.
Agglomerate, rock, 164 ; 167.
—————— description of, 166.
Agricultural Association of Fiji, 174.
—————— native, 74 ; 78 ; 80.
—————— of Government and settlers, defective, 139.
—————— products, 171.
Alligator pear, Avocada, fruit, 99.
Allspice, 106 ; 127.
Alstonia, the Fijian caoutchouc, 195 ; 208.
Alyxia, climber, 196.
Angona, native beverage, 107.
Angora goats, herds, 49 ; 50 ; 191.
Anointing, by natives, 103.
Ants, black, 67.
Apocynáceæ, a natural order of plants, 195.
Apple, Malay, 96.
Aqueducts, native, 76.
Arrowroot plantations, 6 ; 104.
—————— its manufacture by steam, 105.
—————— value of its exports, 106.
Association, Fiji Agricultural, 174.
Asthma, causes of, 143.
Athletic games, 55.
Auckland, New Zealand, vessels from, 57.
—————— distance from Fiji by steam, 186.

B.

Ba, river, 44 ; 49 ; 155.
Babuca, village, 45 ; 183.
Baggage, for travelling, 4 ; 193.
Baker, Mr., 59.

Bakoi, fruit, 97.
Balawa, Pandanus or screw pine, 22 ; 23 ; 49 ; 50 ; 97 ; 110.
Bamboos, 69 ; 125.
Bananas, 37 ; 45 ; 81 ; 93.
Barges, 156.
Bark-cloth-tree, Masi, 13.
Barometer, readings of, 250.
Barrier reefs, 149 ; 154.
Basalt, rock, 164 ; 168 ; 179.
—————— columnar, 165.
Basketwork, 111.
Baths, hot, used as curatives, 164.
Bats, 192.
Bays, rivers fall into, Savu-savu, Natawa, and Sandalwood, 151.
Bêche de mer, 21 ; 193.
Beetles, 193
Benga, 157.
Biba, village, 34.
Bilo, village, 43.
Blacklead, 169.
Boats, native, 55 ; 114.
Boiling springs, 17 ; 163.
Bones of victims, 42.
Bosè vaka Yasana, tribal council, 11 ; 190.
Botanic garden, visit to a, 27.
—————— gardens, 138.
—————— situation and area requisite for, 140.
Bottle-gourd, 97.
Breadfruit, 10 ; 17 ; 25 ; 28 ; 82.
Bread, making of native, 86.
—————— materials made of, 88.
Breccia, rock, 164 ; 179.
—————— description of, 166.
Bridges, native, 25.
"British subject," Kai Biretania, 35.
Bua, province, 50 ; 203.
—————— mountains, 18 ; 151.

Bali, district chief, 10 ; 11.
Barota, district and village, 4 ; 162.
Burying grounds, native, 52 ; 102.
Butterflies, 193.

C.

Cacoa, chocolate, 8 ; 48 ; 52 ; 139, 140 ; 182.
────── area suitable for growth of, 177.
Cakaudrove, province, 10 ; 55.
────── Roko of, 10 ; 54 ; 55.
Calcareous rocks, 164.
────── strata decomposed, 170.
────── soil, 179.
Calms, 144.
Camphor, 127.
Canals, native irrigation, 15 ; 45 ; 76.
Candle nut, *Lauci*, 63 ; 127 ; 187 ; 188.
────── export, value of, 1875, 187-189.
Cannibals, 35 ; 42.
Canoe, native, 13 ; 25 ; 52 ; 55 ; 83 102 ; 114 ; 118 ; 154 ; 156.
Caoutchouc (*see* App. I.), 139 ; 195-202.
Capitalists, 175 ; 177.
Capital of Fiji, 17.
────── site of future, 29.
Capital, want of, 187.
Carica papaya, fruit, 97.
Carriers, 3 ; 4 ; 9-12 ; 17 ; 19.
Cassava, plant, 105.
Caterpillar, destructive, 30 ; 174.
Cattle, herds, 17 ; 30 ; 45 ; 50 ; 51 ; 53.
────── over abundant, 191.
Caverns, 160.
Caves, 42 ; 166.
Cedar tree, *Toon*, 27 ; 135.
Cereals, 74 ; 80.
Channels, deep water, 149.
Chaplain, 41.
Chestnut, Polynesian, *Ivi*, 71 ; 87.
Chinamen, 89.
Churches, construction of, 122.
────── in every village, 1.
Cibicibi, timber tree, 64 ; 120 ; 139.
Cinchona, 46 ; 139 ; 183.

Cinchona, area suitable for growth of, 177.
Cinnamon, 60 ; 106.
Circular letter from Governor to native chiefs, 3.
Clearing for planting sandalwood, 205.
────── for planting of cocoa-nuts, 173.
────── yams, 75.
────── *Dalo*, 76 ; 77.
Cliffs of limestone and of sandstone 166 ; 167.
Climate of Fiji, 142.
────── depressing in summer, 144.
────── healthiness of, 143 ; 186.
Climbers, 110 ; 195 ; 196.
Cloth, native, 109.
Clothing, of natives, 109 ; 110.
────── of Europeans, 144.
Clove, 106 ; 127.
Cocoa-nut plantations, 8 ; 13 ; 14 ; 15 ; 17 ; 18 ; 21 ; 22 ; 23 ; 25-28 ; 50-54 ; 56 ; 60 ; 81 ; 173.
────── value of exports from 1876-1878 ; 171.
────── oil, value of exports, 171.
Coffee, plantations, 6 ; 9 ; 13 ; 25 ; 28 ; 30 ; 31 ; 34 ; 35 ; 37 ; 39 ; 45-48 ; 50-54 ; 139 ; 140 ; 186.
────── area suitable for growth of, 177.
Coir fibre, 8 ; 172.
────── value of exports from 1876-1878, 171.
Coko, ulcerous disease, 143.
Colo ni Suva, village, 31.
Colo, mountain districts, 81.
Commissioner of Lands, 52.
────── of the Governor, 41.
Conservancy of forests, 127 ; 131 189.
Constabulary, 41.
Coolies, Indian, period and terms of engagement, 185.
Copper, 169.
Copra, 8 ; 28 ; 56 ; 171.
────── value of exports from 1876 1878, 171.
Coral limestone, 167 ; 168.
────── reefs, 149.

Cotton, plantations, 6; 7; 13; 28; 32; 49; 54; 170; 171; 181.
—— —— export value from 1875–1877; 180.
—— —— seeds, export value of, 180.
Council of tribe, *Bosè vaka Yasana*, 190.
Crab, land, 193.
Creepers, 62; 195; 196.
Croton-codœium shrub, 27; 102.
Crown grants, 52; 187.
Cultivation, native, 74–92.
—— —— migratory, 62.
Curiosities, 142.
Curtains, native, 109.
Custard apple, fruit, 99.
Cutter, 53.

D.

Dakua, timber tree, 10; 20; 32; 70; 116; 131; 208.
Dakua salu salu, timber tree, 20; 32; 114; 116; 131; 208.
Dalo, 37; 40; 45; 75; 76; 141; 183.
Damanu, timber tree, 20; 32; 37; 48; 71; 114; 115; 208.
Dand's Peak (*Korobato*), 151.
Dawa, kind of litchi, 95.
Deciduous trees, 64; 96; 113.
Deep water, 150.
Deforestation, causes of, 80.
—— —— effects of, 132.
Density of juice of sugar cane, 175; 176.
Devil's Bure, 55; 165.
Diarrhœa, causes of, 143.
Dilo, timber tree, 17; 70; 114; 115.
—— —— oil, 114.
Disintegrated volcanic rocks, 169.
District council, functions of, 11.
Dogs, 192.
Draught oxen, 191.
Drauka, plant, 91.
Drayton's Peak, 151.
Drega guruguru, caoutchouc tree, 195; 198.
Drekiti, river, 20; 151.
Dress, native, 109; 110.

Dress, European, 144.
Dribi, 167.
Drink, of natives, 74; 107.
Drought, cause of, 80; 128; 152; 159.
Drua-drua, island, 22; 152.
Drums, native, 112; 113.
Dry localities, 145.
—— —— dock, 17.
Ducks, 192.
Dyes, 116; 190.
Dysentery, cure for, 99; 143.

E.

Education, of the natives, 3.
—— —— industrial, for the natives, 141.
Elephantiasis, 143.
Engagement, of labourers, 184.
Epidemic, of measles, 81.
Estuaries, 52.
—— —— tidal, 60.
Exotic trees, 49; 135.
—— —— fruits, 97–101.
Exports, 171.

F.

Falls, 175.
Fans, native, 110.
Fawn harbour, 52.
Fecula, 104; 105.
Females, native, hospitality of, 15.
Fences, 98; 126.
Ferns, 9; 20; 45; 46; 48; 51; 58; 59; 64; 65; 119.
—— —— number of, 58.
—— —— tree, 65.
Fertility of soil, 29; 45; 48; 51–53; 93; 186.
Fibres, of banana, 82.
—— —— of cocoa-nut, 172.
Fig tree, 72; 100.
Fighting fence, 94.
Fiji grass, 102.
Fireflies, 193.
Fires, periodical, 7; 16; 20; 30; 40; 42; 46; 49; 51; 132.
—— —— precautions suggested, 133.

Firewood, 121 ; 131.
Fish, 192.
Flies, 193.
Flint, 169.
Floating island, 24.
Floods, 47 ; 128.
Flora, 58-73.
—————— of Polynesia, 58.
Flowering plants, 9 ; 20 ; 48 ; 51 ; 59 ; 102.
—————— number of, 58.
Flying fox, 192.
Fogs, 44 ; 128.
Food, of natives, 74 ; 77 ; 81 ; 86 ; 87 ; 90-92 ; 97 ; 176.
—————— preservation of, 88.
Foreign vessels, port for, 27.
Forests, 30 ; 32 ; 37 ; 40 ; 42 ; 44-48 ; 50 ; 51 ; 54 ; 62 ; 13.
—————— conservancy of, 127 ; 131 ; 189.
—————— destruction of, 38 ; 160.
—————— ordinance, suggestions for, &c. (Appendix III.), 213-237.
—————— restoration of, 127.
—————— rules suggested for regulating falling of timber in, 238-246.
—————— virgin, 61.
Forest creek, 54.
—————— staff, 137.
Fort Carnarvon, 41.
Fortifications, native, 36 ; 98.
Fowls, domestic, 191.
Fringing reefs, 149 ; 154.
Frogs, 193.
Fruit, 93 ; 139.
—————— exotic, 97.
—————— favourable climate for, 99 ; 100.
Fungus, 109 ; 180.
—————— on the skin of natives, 38.
Furniture, articles of, 126.
—————— timber for, 120.

G.

Gale, 53 ; 145.
Gali, 54.
Gardenia, 46 ; 53.

Gardens, 89 ; 90 ; 104.
—————— Botanic, 27 ; 138 ; 140.
Gasau, grass, 126.
Geese, 192.
Ginger, 107.
Gorge, 159.
Government House, 126.
—————— Offices, 126.
Governor's residence, 140.
Gram, 140.
Granadilla, fruit, 99.
Graphite, 169.
Grasses, 68, 102.
—————— nourishing quality of, 191.
—————— scarcity of, 69.
Great council, annual, 12.
Guavas, fruit, 97.
Guides, 3 ; 4.
Gum resin, *Makadre*, 116.
Gumu, acacia richii, 208.

II.

Hale's Peak, 151.
Halting places, of natives, 20.
Harbours, 29 ; 49.
—————— natural, 17 ; 149 ; 161.
Harvey, Dr., 58.
Horses, rearing, 45 ; 50.
Hospitality of natives, 163.
—————— of settlers, 51.
Hot baths, 164.
Hot springs, 163.
—————— used by natives for cooking, 164.
Houses, native, 28 ; 40 ; 52 ; 119.
—————— construction of, 122 ; 125.
—————— of the poor, 124.
Hurricanes, rarity of, 145.

I.

Implements of agriculture, native, 80.
India-rubber, caoutchouc, 195-202.
—————— as prepared by natives, 199.
—————— its market value, 200 ; 202.
Insect, destructive to cocoa-nut tree, 30 ; 174.
—————— stick and leaf, 193.
Institute, Mechanics, at Levuka, 141.

Interpreter, 4 ; 22 ; 39 ; 208.
Iron, 169.
Islands, number of in the Fiji group, 149.
────── elevation of, 150.
────── estimated area of, 149.
────── situation of, 149.
────── physical character of, 150.
Ivi, tree, 17.
────── interesting, 71.
────── nuts, 87 ; 88.

J.

Jack, fruit tree, 183 ; 100.
Juice of sugar cane, density of, 175, 176.
Justice, administration of, by magistrates, 21.

K.

Kalakakaisau, fibre-yielding plant, 111.
Kali Kosa, village, 24.
Kaluba, village and river, 30 ; 94 ; 166.
Kami Rusai, village, 47.
Kanacéa, island, 26.
Kadavu, island, 35.
Kau Karo, cure for ringworm, 38.
Kau Kuru, timber tree, 63 ; 118.
Kawrie pine, Dakua, timber tree, 116.
Kausia, tree, 20 ; 117.
Kauta, sedge, 110.
Kau-tabua, tree, 20 ; 32 ; 117.
Kava, native beverage, 107.
Kavika, Malay apple, 96.
Kiou, island, 167.
Kobalau point, 53.
Koka, timber tree, 119.
Koro, island, 26 ; 105 ; 161 ; 169.
──────, native village or town, 11 ; 37 ; 50 ; 206.
────── tidily kept, 21.
Koroba, mountain, 158 ; 208.
────── Pickering's Peak, 42.
Koro-i-vono, town, 9.
Koro Levu, village, 52.
Koro Loa, mountain, 32.
Koro ni Saca, town, 12.

Koro Suli, village, 46 ; 156.
Koro Wai-wai, 33.
Kulava, timber tree, 121.
Kuru Kuru, river, 23.

L.

Labasine, village, 22.
Labour, distribution of, 79.
────── imported, 184.
────── native, 185.
────── supply of, 139.
Labourers, agricultural, their character ; habits ; terms of engagement ; wages ; passage ; food, &c., 184.
Lakes, not numerous, 154.
────── salt, 153.
Laucala, island, 26.
Land crabs, 193.
Landslips, 40 ; 128 ; 208.
Langton, Mr., 31 ; 208.
Language, diversities of, 39.
Lasalasi, village, 34.
Latitude, 247.
Lavoni valley, 4 ; 162 ; 169
Leaf insects, 193.
Lemon grass, 69.
────── trees, 42.
────── fruit, 93 ; 95.
Levelling, by natives, 77.
Levuka, present Capital of Fiji, 26 ; 143 ; 167.
────── objectionable situation of, 162.
Lewininini, timber tree, 20 ; 117 ; 208.
Liberian coffee, 8 ; 9 ; 48 ; 51-53.
Library at Levuka, 141.
Liku, native dress, 109.
Limes, 93.
Limejuice, 95.
Limestone rock, 164 ; 165 ; 177.
────── cliffs, 166.
────── disentegrating, 208.
Litchi, longan, fruit trees, 100.
Loam, 15.
Lobsters, 193.
Localities, dry and wet, 145.
Loma loma, village, 19 ; 27 ; 161 ; 167.

Loma loma, island, or *Vanua Balavu*, 'exploring islands,' 26.
Longevity of natives, 143.
Longitude, 247.
Loquat, Bibassier, 98.
Lose lose, fruit, 97.
Lyonsia, climber, 196.

M.

Macou, 106.
Macuata, province, 19 ; 203.
Magistrates administering justice, 21.
——— native, 11.
Mahogany tree, 27 ; 136.
Mail (steam) service, 26 ; 49 ; 57.
Maize, 49 ; 50 ; 54 ; 170 ; 207 ; 188.
Makadre, gum resin, 116.
Malachite, 169.
Malaka, or Malata, village, 24.
Malarial fever unknown, 142.
Male, nutmeg, 106.
Mali, island, 22 ; 151.
Mango, fruit, 99.
Mango, island, 26 ; 28 ; 169.
Mangosteen, fruit tree, 100.
Mangrove, *tiri*, tree, 7 ; 17 ; 22 ; 23 ; 50.
Mantis, 193.
Markets for produce, 184 ; 186.
Marl, rock, 164.
Maro, 182.
Marshes, 45.
——— salt water, 60.
Masi, Bark-cloth tree, 13.
——— as used by natives, 108.
Mataiavai, plant, 111.
Matawala, village, 43.
Matting, 110 ; 172 ; 193.
Mauritius fever, 33.
Measles, epidemic of, 31.
Mechanics' Institute at Levuka, 141.
Meke, native dance or play, 55.
Melbourne, steamer from, 57.
——— distance from Fiji by steam, 186.
Meteorology (*see* App. V.) 147 ; 247 ; 255.
Meteorological observations. (*see* App. V.) 247 255.

Milne, 58.
Model farm, 140.
Mokogai, island, 56.
Molasses, 174.
Moli Kana, chaddock, 93.
Moli Kuru Kuru, lemon, 93 ; 94.
Moli ni Tahiti, orange, 94.
Monsoon, 145.
Mosquitoes, 24.
——— native curtains for, 109 ; 193.
Mosses, 61.
Moth, destructive, 30 ; 174.
Moturiki, island, 5.
Mountains, 18 ; 32 ; 35 ; 46-48 ; 50 ; 60.
——— their elevation, 150 ; 152 ; 153 ; 160 ; 161.
Mud flats, 50 ; 51.
Mulberry, fruit, 98.
Mulo-Mulo, timber tree, 118.
Museum, utility of, 142.
Mystical rites, 165.
Mythological stories, of natives, 24.

N.

Na Babuca, 45 ; 83.
Na Bekè Lukè, 47 ; 156.
Na-Colo-Suva, town, 31 ; 140.
Nadi, 49 ; 50 ; 158.
Nadoya, 24 ; 193.
Nadrau, village, 43 ; 183.
Nadroga, harbour, 49.
Naduri, village, 19.
Na Koro Vatu, village, 5 ; 6 ; 46.
Naloa, island, 151.
Namatu falls, 157.
Namena, 158.
Na Moali, village, 39.
Na Mouli, 183.
Na Moali, 40.
Namosi, village, 32 ; 94 ; 135 ; 159 ; 169.
Nananu, island, 49 ; 50 ; 158.
Na Quara-wai, 156 ; 157 ; 169.
Na Quave, village, 31.
Na Sali Levu, 54.
Na Saucoka, 90 ; 208.
Nasova, Government House at, &c., 126.
Nataimba, island, 27.

Natawa bay, 10 ; 13 ; 55 ; 163.
Natives, instincts of, 74.
———— ingenuity of, 77 ; 78 ; 80 ; 109.
———— their taste for flowers, 102.
———— healthiness and longevity of, 143.
———— habits of, 144 ; 190.
———— hospitality of, 163.
Native dance, *meke*, 55.
Native preachers, 4 ; 41.
Na Tuatuacoko, Fort Carnarvon, 40 ; 158 ; 208.
Na Saucoko, village, 42 ; 208.
Navaloa, 140 ; 6.
Navatu island, 193.
Navigable streams, 23 ; 51 ; 152.
Navigation, facilities and difficulties of, 149 ; 161.
Navosa, province, 40 ; 134 ; 203 ; 205 ; 189.
———— quantity of sandalwood growing in, 211.
Navua, river, 29 ; 49 ; 157.
Nawanawa, fruit, 97.
Nawasakubu, town, 208.
Needle Peak, 151.
Nets for fishing, 111.
Noko noko, casuarina, timber tree, 23 ; 120.
Nukusari, village, 32.
Nursery, botanic, 139.
Nutmeg, *Male*, 106 ; 127.
Nuvera island, 151.

O.

Occupation of natives, 111.
Oil, cocoa-nut, value of, 171.
Oncocarpus, tree, 38.
Ophthalmia, of natives and settlers, 143.
Orange, 93 ; 94.
Orange cowry, 193.
Orchids, 66 ; 119.
Ornamental plants, 102.
Ornaments of natives, 103.
Ovalau, island, 31.

Ovalau, its central position, 161.
———— its scenery, 162.
Oxen, draught, 191
Oysters, 193.

P.

Palms, 61 ; 62 ; 68.
Papalagai, foreigner, 24 ; 185.
Parrots, 64 ; 192.
Pasture, 45 ; 50.
———— destruction of, 59.
Patent, 190.
Paving stones used by natives, 166.
Peach, fruit, 100.
Peaks, 151.
Pearl shell, 193.
———— export of, 193.
Pepper, plant, 107 ; 108.
Perfumes, native, 103.
Pests, 59 ; 180.
Pickering's Peak, *Koroba*, 42 ; 158 ; 169.
Pigs, 15 ; 126.
———— whence imported, 191.
Pine apple, *Balawa ni papalagi*, foreign Pandanus, 93.
Pitcher, bamboo, 125.
Plantations, *see* sugar cane, &c., 171–183.
Planting of yams, 75.
———— of Dalo, 76.
———— according to Fijian mythology, 97.
———— for climate purposes, 207.
Plants, ornamental, 102.
———— distribution of to inhabitants, 139.
Pleasure ground, 140.
Ploughing, 175.
Polity, tribal, past and present, 11.
Pomegranate, fruit, 99.
Population, diminution of, 31.
Porphyry, rock, 164.
Porpoises, 192.
Port for foreign vessels, 27.
Potatoes, 88.
Pounac, food for cattle, 172
Prawns, 193.

Produce, European, 89.
——— markets, 186.
Products, agricultural, 171.
——— by the natives, 187.
Psidium cattleyanum, 97.
——— chinensis, 97.
——— pomiferum, 97.
Pulping of coffee, 177.

Q.

Qele Levu, island, 193.
Quame, island, 26.

R.

Rabi, (Rambi), island, 8; 26; 161; 163; 167; 169.
Rafts of bamboos, 34; 157.
Rainfall (see App. V.) 44; 46; 51; 61; 128; 145; 159; 160; 247-255.
Raki Raki, 155, 157.
Rapids, 157.
Rara, or Square, village, 56.
Raspberry, fruit, 100.
Ratan, 111.
Ratoons of sugar cane, 175; 176.
Rats, 193.
Raurau, bay, 158.
Ravines, 44; 51; 63; 159; 208.
Rebellion, 41.
Reefs, coral, 149.
Reforesting, 127; 128; 134; 204.
Regulations (suggested) for planting, &c., of sandalwood, 209.
——— for felling timber in Government forests, 238; 245.
Religious observances, 3.
Rerega, turmeric, 105.
Reserves, forests, 215; 223; 227; 238.
Rewa, river, 5; 29; 31; 94; 155.
Re-wooding, 127; 128; 134; 204.
Rheumatism, native cure for, 114.
———.——— causes of, 143.
Rice, 140; 183; 186.
Ringworm, 38.
Rivers, 150.
——— navigable, 152
——— their size and beauty, 154; 155.

Road making, native, 23.
Roadsteads, 149; 160.
Rocks, 164.
——— influenced by volcanic heat, 165.
Roko Tui, supreme chief of a province, 10.
Rope, 111; 172.
Route through Fiji, 3.
Rukuruku village, 46.
Rum, 174.

S.

Sagali, timber tree, 118.
Sago palms, 63; 68.
Sailors, native, 55.
Salt lake, 153; 169.
Sandalwood, 22; 42; 50; 63; 203.
——— season for planting, 205.
——— its market value, 204.
——— transplanting of, 209.
——— tabu, 209.
Sandstone, 164.
——— cliffs, 167.
San Francisco, mail viâ, 57.
Sau-sau, passage, 152.
Savages, imported, 184.
Savoo, timber tree, 121.
Savu-savu, bay, 10; 16; 52; 163; 169.
Saw mills, 30.
Scenery, beauties of, &c., 29; 31; 34; 35; 40; 46; 47; 52; 153; 160; 162.
Schools, construction of, 122; 141.
——— industrial, 140.
——— training, at Navaloa, 140.
Scoria, 168; 169; 179.
Scott and Harvey's settlement, 157.
Screens, native, 109.
Sea, fruit, 97.
Sea Island cotton, 181.
Sea-shore plants, 59.
Seasons, dry and wet, hot and cool, 144.
Sedge, 69.
Seeds, supply of, 139.
——— distributed by government, 179.
——— danger attending importation of, 180.

Seeds, export of (cotton) 180.
Seemann, Dr., 58.
Serua, 49; 157.
Settlers, 18; 22; 27-31; 49-54; 93; 97; 138; 143; 187.
Shaddock, tree, fruit, 35; 93; 95.
Shale, 164.
Shallow water, 150.
Sharks, 23; 47; 192.
Sheep, 45; 50; 56; 191.
Shells fish, 193.
Shell, 193.
Siga Toka, river, 40; 42.
——————— peculiarities of its course, 158; 163.
Silk, 99.
Silkworms, 50; 98.
Sinnet, coir, 111.
——————— native industry for ornaments, 124; 126; 172.
Site of future capital of Fiji, 29.
Sites of native houses, 124.
Smith and Aitchinson's, 54.
Snakes, 192.
Soapstone, 166.
Soil, fertility of 29; 45; 48; 51-53; 93.
——————— scantiness of, 60.
——————— composition of, 69; 170; 179.
Soldiers, native, 41.
Somosomo, *Taviuni*, village, 54.
Sores, causes of, 143.
Sources of streams, 150.
Specimens of plants, 18; 23; 32; 36; 37; 45; 46; 48; 51; 58; 59.
——————— difficulty of preserving, 19.
——————— total number collected, 19.
Spice trees, 107; 139.
Squalls, 145.
Steamer (mail) 26; 49; 93.
Steam Navigation Co., Australasian, 57.
——————— Peninsular and Oriental, 57.
Stick insects, 193.
Stock-raising, 31; 45; 49; 50; 54; 191.
Strangers' house, *Bure ni sa*, 12.
Strawberry, fruit, 100.
Strata, 44; 46; 170.
——————— results of doubling up of, 165.

Streams, 150.
——————— navigable, 150.
——————— complicated sources of, 150.
——————— abundance of, 177.
Sugar, 186.
——————— cane, varieties of, 174.
——————— its adaptation to the soil, 5-9; 12-15; 20-25; 30-31; 35; 43; 47 -53; 54; 69; 139; 140; 170; 175.
——————— disease, dangers of, 129.
——————— manufacture of, as used by the natives, 174.
——————— export value, 1876-1878; 174.
——————— area of land suitable for growth of, 175.
Sugar Loaf, mountain, 151.
Sugar mills, want of, &c., 5; 6; 18; 23; 31; 49; 54; 174; 175; 186.
Suggested regulations for planting sandalwood, 209;
Sulu, native cloth, 109.
Suva, chosen site of capital of Fiji, 17; 29; 140.
Swampdalo, 183.
Sweet potatoes, *Kumara*, 84.
——————— price of, 84.
Sydney, mail from, 57.
——————— distance from Fiji by steam, 186.

T.

Tabernæmontana Pacifica, tree, 195.
Tabu, forbidden, 81.
——————— native observance of, 110.
Tagaloa, district, 10.
Tai Levu, district, 6.
Takia, dug out canoe, 22, 118.
Tamarind, fruit, 100.
Tamavua, river and village, 30; 31; 166.
Tapioca plantations, 6; 87.
Tarawan, fruit tree, 96
Taro, 15; 183.
Taukuku ni vuaka, weed, 91.
Taviuni, island, 10; 26; 160.
——————— of purely volcanic origin, 168.
Tavola, timber tree, 113.

Tavua, river, 49.
——— Peak, 50.
Tax, Inspector of, native, 203; 239; 240; 241; 242; 243; 244; 245.
Taxes, of natives, 182.
——— of natives, in kind, 187.
Tea, 46; 139; 140; 182.
——— area suitable for growth of, 177; 186.
Temperature (*see* App. V.) 147; 247-255.
Temple, heathen, 45.
Terraces, native, 77.
Thatch of houses, 123; 126.
Thermometer, readings of, 247-255.
Thunderstorms, 22; 43; 144.
Timber industry, 30; 38; 42; 112; 139.
——— importance of growing it, 131.
——— kinds recommended, 135.
——— trees, 112-126.
Tinika, reed, 55.
——— throwing the, 56.
Title deeds (land), 52; 187.
Tobacco plantations, 49; 170; 181; 187; 188.
Tokalu, Solomon Island ringworm, 38.
Tonga, island, king of, men of, 27.
——— ——— pigs from, 191.
Tools, 141.
Toon, cedar tree, 27; 135.
Torches, travelling with, 24.
Trachyte, rock, 164.
Trade winds, 144.
Transplanting of sandalwood, 209.
——— of coffee, 179.
Travellers, advice to, 193.
Travelling difficulties of, 20, 23-25; 32; 33; 39; 40; 43; 46; 48; 51; 53.
Trees, size of, 70.
Tribes, native, 42.
Tufa, 168; 169; 179.
Turaga ni Koro, village chief, 11; 12; 22.
Turaga ni lawa, native magistrate, 11.
Turban, native, 109.
Turkeys, 192.
Turmeric, plant, 63; 105.

Turtle, 193.
Tutu, 22; 152.

U.

Udu point, 203.
Ugavule, land crab, 193.
Ulcerous disease, *coko*, 143.
Upheaval of coast, 154.
——— results of, 165.

V.

Vagadaea, 167.
Vai vai, timber tree, 118.
Vanilla, 182.
Vanua Leva, island, 9; 14; 19; 55; 163; 164.
Vatu Kali, 53.
Vatu Kura, 25.
Vau, plant, 110.
Vegetable débris, 169.
Vegetables, European, 89.
——— Fijian, 90; 91.
Vegetation, 61.
Veseri, village and river, 48.
Vesi, timber tree, 17; 112.
Vieue, village, 13.
Vienunga, 45.
Vienunga, village, 35; 37.
Village, *Koro*, 11.
Vine, 100.
Viti Levu, island, 5; 29; 164; 179; 189.
Voivoi, plant, 110.
Volcanic heat, its influence on rocks, 165.
Volcanoes, extinct, 154; 160; 168.
Voma, lofty mountain, 159; 164.
Vosi Dam, village, 40.
Vuga, timber tree, 42; 119; 208.
Vuga vuga, timber tree, 119.
Vuna point, village, 26.
Vuni Sawani, village, 13.
Vuni vutu, village, 22; 152.
Vutukana, fruit, 97.
Vutu, tree, 70.

W.

Wagodra-godra, wild bramble, 97.
Wai Bas-aga, village, 40 ; 163.
Wai Delici, river, 7 ; 155.
Wai Dini, river, 34 ; 156 ; 179.
Wai Dradra, village, 45.
Wai Levu, district, 151.
Wai Levu, river, 18.
Wai Manu, district, 29.
——— river, 48 ; 156.
Wai-ni Awa, river and village, 36.
Wai ni Buka, river, 47 ; 155.
Wai ni Buli, Tasman's straits, 26.
Wai ni Koro, stream, 24.
Wai ni Loa, river, 45 ; 46 ; 56 ; 165.
Wai ni Mala, river, 45 ; 156 ; 179.
Wainunu, district, 51.
Waiwai, 18.
Wa Loa, fungus, 109.
War, between natives, 42.
——— effects of, 83.
——— clubs, native, 121.
Warei, village, 51.
Wasa Kuba, village, 169.
Water melon, 97.

Watersheds, 150 ; 151 ; 158.
——— peculiarities of, 150 ; 151 ; 158.
Weeding, 187.
Weeds, 59 ; 76 ; 91.
Wesleyan Missions, 3.
——— — training school, 6.
Wet localities, 145.
Whales, 192.
Wi, fruit tree, 96.
Wild ducks, fowl, pigs, 192.
Wild sugar-cane, 69.
Wind, direction and force of, 247–255.
Winds trade, variable, &c., 144 ; 159.
Women, native, 79.
Woods and Forests Committee, 136.
Wreaths, native, 103.

Y.

Ya, sedge, 110.
Yabia, plant, 104.
Yaka, plant, 111.
Yams, 13 ; 34 ; 37 ; 45 ; 74 ; 75 ; 86 ; 141.

LONDON:
Printed by GEORGE E. EYRE and WILLIAM SPOTTISWOODE,
Printers to the Queen's most Excellent Majesty.
For Her Majesty's Stationery Office.

[P 456.—750.—5/81.]

www.ingramcontent.com/pod-product-compliance
Lightning Source LLC
Chambersburg PA
CBHW021956220426

43663CB00007B/842